Housing, Care and Inheritance

Housing and society series
Edited by Ray Forrest
School for Policy Studies, University of Bristol

This series aims to situate housing within its wider social, political and economic context at both national and international level. In doing so it will draw on the full range of social science disciplines and on mainstream debate on the nature of contemporary social change. The books are intended to appeal to an international academic audience as well as to practitioners and policymakers – to be theoretically informed and policy relevant.

Housing and Social Policy
Contemporary themes and critical perspectives
Edited by Peter Somerville with Nigel Sprigings

Housing and Social Change
East–West perspectives
Edited by Ray Forrest and James Lee

Urban Poverty, Housing and Social Change in China
Ya Ping Wang

Gentrification in a Global Context
Edited by Rowland Atkinson and Gary Bridge

Housing and Social Transition in Japan
Edited by Yosuke Hirayama and Richard Ronald

Forthcoming:

Sustainable Development
A new perspective for housing analysis
Rebecca Chiu

Housing, Care and Inheritance

Misa Izuhara

LONDON AND NEW YORK

First published 2009
by Routledge
2 Park Square, Milton Park, Abingdon, Oxfordshire OX14 4RN

Simultaneously published in the USA and Canada
by Routledge
711 Third Avenue, New York, NY 10017

First issued in paperback 2016

Routledge is an imprint of the Taylor & Francis Group, an informa business

© 2009 Misa Izuhara

Typeset in Times and Frutiger by
Keyword Group Ltd, Wallington, Surrey, UK

All rights reserved. No part of this book may be reprinted or reproduced or utilized in any form or by any electronic, mechanical, or other means, now known or hereafter invented, including photocopying and recording, or in any information storage or retrieval system, without permission in writing from the publishers.

The publisher makes no representation, express or implied, with regard to the accuracy of the information contained in this book and cannot accept any legal responsibility or liability for any efforts or omissions that may be made.

British Library Cataloguing in Publication Data
A catalogue record for this book is available from the British Library

Library of Congress Cataloging-in-Publication Data
Izuhara, Misa.
  Housing, care and inheritance / Misa Izuhara
    p. cm. — (Housing and society)
  Includes bibliographical references and index.
  1. Older people—Housing. 2. Older people—Care. 3. Older people—Family relationships. 4. Inheritance and succession. I. Title.
  HD7287.9.I98 2008
    362.61—dc22   2008013908

ISBN 13: 978-1-138-99175-0 (pbk)
ISBN 13: 978-0-415-41548-4 (hbk)

Contents

List of figures and table vi
Acknowledgements vii

1 Introduction 1
2 The cultural practice of intergenerational reciprocity 8
3 Housing assets and intergenerational transfer in a global context 26
4 Long-term care and shifting state–family boundaries 47
5 Comparative analysis of housing wealth accumulation and family relations 65
6 A comparative analysis of perspectives on inheritance 87
7 Rethinking the 'generational contract' between care and inheritance 110
8 Conclusion 131

References 137
Index 152

Figures and table

Figures

2.1	Cross-national comparison speed of ageing, 7–14 per cent	19
3.1	Residential land prices in Japan and England (1986–2000)	33
4.1	Public and private expenditure on long-term care as a percentage of GDP (2000)	59
4.2	Home care versus institutional care – public expenditure on long-term care as a percentage of GDP (2000)	61
5.1	Attitudes of condominium owners in Japan	74
5.2	Home ownership rates by age of head of household, 1998	82
6.1	Type of households among those aged 65 and over in Japan	91
6.2	Value of asset per household by age group of household in Japan	101

Table

4.1	Share of very-old persons (80+) among older people in selected OECD countries, 1960–2040	50

Acknowledgements

The research which this book is based on is predominantly derived from two research projects funded by the UK's Economic and Social Research Council (R000223717 & RES-000-22-0798) on the 'Generational Contract between Care and Inheritance in Britain and Japan'.

First of all, I would like to thank those families in both Britain and Japan who participated in the above research projects. Comments from Ray Forrest (the series editor), Liz Lloyd and Karen Rowlingson helped to reshape the book and I am grateful to them. I would also like to thank Dawn Rushen for copy-editing the manuscript. Finally, my appreciation has to go to my mother, Sachiko Izuhara, in Japan. She has been my inspiration and a great source of support. Without her networks and child care support, the research projects would never have been completed. I am very grateful to her.

The interpretation of the data and views expressed in this book are entirely those of the author.

1 Introduction

With socioeconomic and demographic changes taking place in contemporary societies, relations between generations within the family have become more diverse and have moved away from conventional ideology and practices. Unlike the social contract mediated by the state, the micro-level generational contract usually involves a direct exchange of goods and services within the family across generations. The norm of reciprocity is one of the significant characteristics of this generational contract, which governs how individuals accept and provide support involving rights and responsibilities, and credits and debts. However, at the micro-level (interpersonal relations), reciprocal arrangements alone do not dictate support provision. The nature of the caring relationship rests on a 'delicate balance between reciprocity, affection and duty' (see, for example, Marshall *et al.* 1987; Qureshi and Walker 1989).

This book explores in particular the connection between housing assets, care provision and inheritance perspectives within the family by locating such interpersonal relations between generations in the wider social, economic and institutional context. In this context, it examines the myth and the changing patterns of the particular exchange of long-term care and housing assets between older parents and their adult children cross-nationally in the East and the West. For example, traditionally, a vital exchange model among Japanese families was that the eldest son (and his wife) had a duty to succeed the family and to provide care to their parents in co-residency, and in return they inherited the entire family wealth; while in Britain, with its established welfare system and cultural norms giving preference to a high degree of independence, care and inheritance do not necessarily go together. Given the different national contexts in relation to cultural norms, family tradition, laws and institutions and housing markets, such micro-level generational contracts are negotiated differently in different societies and in different periods of time.

Recent social change has started to alter the existing patterns of the generational contract. Despite the popular discourse regarding filial piety in East Asian culture, the attitudes of younger people towards the provision of care for their

Introduction

older parents are changing. According to the Sixth International Comparative Survey on the Attitudes of Youth in 1999, for example, only 25 per cent of Japanese youth aged between 18 and 24 answered that 'they would take care of their parents by all means', while more than twice as many respondents (66 per cent) mentioned 'they would rather provide care within the limits of their physical and financial ability'. However, the same survey revealed more willingness among the young in the UK towards family support (50 per cent and 44 per cent, respectively). The survey in various European countries also found that filial norms were more prevalent among younger than older people in England (Daatland and Herlofson 2003). Although this may be entirely perceptional rather than practice-based, the findings are somewhat puzzling, given the common understanding of social norms in the East (familism) and the West (individualism). This book will therefore paint a picture of changing attitudes and practice towards support provision across generations through identifying various key factors influencing such changes.

Comparative policy logic is another analytical framework adopted in this book, examining in particular changing patterns of long-term care provision in the family and society. The micro-level generational contract does not exist on its own but is influenced by the more macro social structure and public policy. It is particularly the case in societies with an established but shifting welfare state. Policies on welfare provision tend to draw a boundary between the state and the family, but such a boundary is highly contested and shifts over time. With the power of law and public policies, the state is often a facilitator in drawing such boundaries by encouraging or discouraging certain family behaviour. At the same time, the state can also be a follower of trends and implement some measures to support and confirm already shifted boundaries. How policies deal with long-term care issues is indeed a central concern of an ageing society.

'Who is responsible for the cost and delivery of care' has been well debated in policy and politics in advanced welfare states in recent years, and we now find a number of shared policy directions globally, including: a home care-centred approach; a mixed economy approach by expanding alternative providers from the conventional family–state nexus; and expanding care options to include cash benefits to informal carers. Writing this introduction in 2008, welfare reform in relation to older people (and people with a disability) is moving towards a cash-for-care approach where users purchase care rather than receive traditional service provision (see, for example, Arksey and Glendinning 2007; Ungerson and Yeandle 2007). The level of choice that users can exercise, however, varies significantly between societies and jurisdictions within societies. Issues still remain in areas such as how to overcome existing barriers for individuals to fully exercise their choice and how policy makers ensure the accountability in the use of public funds. In the recent policy reform under the social insurance on long-term care, Japan has also achieved commodification and wider socialization of care beyond the conventional family practice. Whether or not the new scheme would reward family carers was extensively debated in the planning process and the decision not to offer a cash payment to family carers made the Japanese scheme different from the German model (Masuda 2002). In this context, the policy reform in Japan provides an interesting case study and the book explores whether the new policy directions reflect or reinforce the patterns of informal family care.

The third dimension of the book is an asset-based approach in welfare, considering home ownership as an exchangeable commodity for individuals and families. The debate on asset-based approaches in public policy originated in the United States and has been transported to many post-industrial economies in the West including the United Kingdom, Sweden and Australia. This approach views inequalities in the ownership of assets as problematic and tries to alleviate such inequalities by helping those on low incomes to accumulate assets through policy measures such as individual development account schemes (Regan and Paxton 2001). Sherraden (1991) also emphasizes the importance of assets for low-income households as

> income only maintains consumption, but assets change the way people interact with the world. With assets, people begin to think for the long term and pursue long-term goals. In other words, while income feeds people's stomachs, assets change their minds
>
> (1991: 13)

The important role that assets play in contemporary society is to allow individuals and households access to security, independence and opportunities on top of the mere ownership of financial and material assets.

In many capitalist societies, accumulation of individual assets occurs largely through institutionalized mechanisms, primarily via home ownership and pension schemes that are clearly defined and heavily subsidized within public policy (Regan and Paxton 2001). Asset-building policies, including promoting home ownership, are primarily geared toward the affluent and have inevitably created social stratification between households. Also, housing often forms the largest part of households' assets, and the increasing rates of home ownership and housing market inflation in some societies has meant the increasing potential for individuals and households to use equity invested in housing throughout their lifecourse. This is highly relevant to the debate about the consequences of societal ageing linking issues of long-term care and an individual's asset accumulation. With the increased economic independence of older people with housing assets, savings and the safety net of social security, older parents could use their accumulated wealth in later life to purchase alternative services in the market, to negotiate support with younger family members and so on. The different ways in which states perceive and treat individual asset accumulation regarding

Introduction

long-term care are, however, manifest in fundamental differences across societies (Izuhara 2005). Overall, this book poses a key question as to whether such a link between care and inheritance would form part of the generational contract in increasingly contested family and social contexts.

Highly international and comparative in perspective, this book addresses important sociological as well as policy questions regarding intergenerational relations involving housing, care and inheritance. It also offers new empirical and theoretical insights into the interconnection between personal housing wealth and family relations cross-nationally. This is a monograph partly based on first-hand empirical research and reflection stretching over a number of years.

Contributions to knowledge and understanding

The strength of this book lies in the materials collected and analysed by the author whose research experiences and insights are embedded in the contrasting cultures and policy contexts of two societies in the East and the West. Although the empirical data are drawn from comparative research between Britain and Japan, this book offers a wider analysis of relevant theories and issues with the international perspectives.

There is currently a lack of empirical data on inheritance and family relations in the academic literature. In Britain, Finch and her colleagues conducted major research based on a set of wills in the early 1990s, and data from the Inland Revenue were also often used to reveal patterns of asset accumulation and transfer (see, for example, Finch *et al.* 1996; Munro 1988; Hamnett *et al.* 1991; Finch and Mason 2000). More recently, the Joseph Rowntree Foundation commissioned a social survey to investigate people's attitudes towards inheritance in Britain (Rowlingson and McKay 2005). In Japan, the issues have been explored using mainly multiple-choice survey questionnaires (see, for example, Noguchi *et al.* 1988; Tokyo Women's Foundation, 1997; Horioka *et al.* 1998). However, considering the complex nature of the intergenerational relations and increasing diversity within families, as well as the sensitivity in relation to household assets and personal finance, this study instead chose a qualitative method of inquiry to address the issues, to expand academic debate in this field, and to fill the gap in the existing literature.

Furthermore, existing comparative research in this area is often focused on European and other Western countries (see, for example, Twigg and Grand 1998). Thus the studies comparing East and the West are relatively rare. As there is a growing interest backed up by a more recent and growing body of literature on East Asian welfare societies (see, for example, Walker and Wong 2005; Takegawa and Lee 2006), an interesting contrast can be drawn comparing family practice and policy intervention between the East and the West. The book also joins a growing body of comparative publications with reference to East–West policy

Introduction

and housing studies. However, many existing comparative publications tend to be a collection of individual chapters contributed by different scholars whose expertise is located in their own society, and thus each chapter tends not to engage with East–West comparisons (see, for example, Alcock and Craig 2001; Forrest and Lee 2003; Izuhara 2003a; Groves *et al.* 2007). In effect this means that the burden of comparison rests largely with the reader. This single-authored monograph instead offers comparative analysis of the chosen themes throughout the volume.

Moreover, Japan has an established but somewhat different welfare regime and well-practised family support and has been experiencing a prolonged recession which has impacted on the housing market. As it involves analysis of Japanese welfare society, this study will thus provide a new perspective and offer different policy comparisons to conventional European and Anglo-Saxon approaches. In addition, cross-national work, particularly among non-English-speaking countries, often depends on aggregated data and has tended to emphasize convergence in social policy development across nations (Kennett 2001). Using a qualitative method of inquiry instead, this study will recognize both convergence and divergence in terms of the experiences of the households and policy development, and is sensitive to national and cultural specificity.

Although data for the UK or Great Britain are often referred to when discussing general trends in this book, most of the interpretative comparisons such as policy areas, in particular long-term care, empirical case studies and inheritance practices, are with England. This is largely because the different parts of the UK have, to a varying extent, separate legal traditions and policies. Also in the context of a multicultural society, it is difficult to reflect the diversity of family traditions and kinship now found in the British Isles. The empirical discussion is thus particular to the kinship forms in the White English population.

Methods and approach

The empirical data for this book are drawn from a series of cross-national, comparative studies conducted in Britain and Japan between 2002 and 2005 (both funded by the Social and Economic Research Council in the UK: R000223717 and RES-000-22-0798). It was based on family ethnographies, composed of stories from two generations of the same family regarding their housing history and choices, inheritance perspectives, as well as formal and informal care. Unlike some other studies on intergenerational relations, in which multiple generations were interviewed simultaneously (see, for example, Brannen *et al.* 2004; Wade and Smart 2004), the research was conducted in two phases. The initial fieldwork was conducted among older people in 2002–3, and subsequent research explored the views and attitudes of the younger generation (the adult children of older informants) in 2005. A simultaneous approach has its strengths, for example, in not allowing family members to influence each other's views, and providing

Introduction

greater confidentiality for each individual interview. In addition, samples would not include any wastage, since only families in which all the relevant members agree to participate were actually interviewed.

Initially, a series of face-to-face interviews with 28 older people in each society were conducted in 2002 and 2003. Purposive sampling was used to select older informants according to their age; the availability of children; and being (or having been) a home owner. Targeting only home owners helped control social class to a certain extent; however, the research included a wide range of income and asset groups. The research found that not only 'age' but also 'which stage of life-course they were at' provided an interesting indicator. For example, the experiences and perceptions of those who had many great-grand children appeared to be very different from those whose daughters had just had their first baby. The research included both men and women with varying marital status and family composition, reflecting their assumed differences of expectations and experiences. In terms of ethnicity, only Japanese and White British elders were interviewed in this study.

The subsequent research among adult children of the previous informants was conducted in 2005. Sixteen Japanese and 14 British adult children participated in the research. In some 'families' (three in Japan and four in Britain), two adult children (siblings) were interviewed from the same family. This helped enhance the analysis of the complex and dynamic nature of family relations by including more than one child of varying genders, family and work circumstances, and proximity to their parents. Contact with prospective informants was facilitated by the older informants from the first study. This may, however, have created a certain bias towards the selection process, since the parents inevitably screened the potential informants and only children who were more available (often women) and had a 'reasonable' relationship with them may have been introduced. Only three out of 14 British informants and four out of 16 Japanese informants were male. Japanese samples included three single co-resident daughters (who provided fascinating accounts of their care duty, economic dependency and housing destinations), while the British research captured many conventional nuclear families consisting of a married couple with dependent children. The research also included some complex and reconstituted families after divorce and re-partnering. Only cases where both parents and adult children agreed to participate were interviewed.

There was a difficulty in drawing a sufficient number of respondents from the existing (already small) pool of samples, and this has proven to be a drawback of such an incremental approach. Not every informant from the first research was eligible to produce subsequent informants. For example, those who had already sold their housing and moved to purpose-built housing for older people were excluded from the subsequent research. To boost the number of informants, the research recruited four new pairs of informants in England and three in Japan, but the number of new recruits was kept to a minimum to retain the connection with the original research.

Structure of the book

This book draws on both theoretical perspectives and empirical analysis regarding the interaction between housing, care and inheritance. It has two parts. The first part (Chapters 2 to 4) reviews key theoretical, conceptual and policy debates in the fields with global perspectives. Chapter 2 explores a theoretical understanding of the reciprocal dimension of intergenerational relations. It examines how family reciprocity is defined and practised differently in different cultures and how such practices have been shifting in the context of contemporary social change. Chapter 3 then provides the global housing context where the two contrasting housing markets of Britain and Japan are located. A review of the growth of home ownership and thus the accumulation of personal housing assets in the family will help understand both the importance and the impacts of inheritance and asset transfer on family relations. This is followed by a chapter on the policy analysis of financing and delivery of long-term care considering trends and developments across OECD (Organization for Economic Co-operation and Development) countries as well as in East Asia. The analytical framework used for Chapter 4 is the relationship between the family and the state. The chapter examines how, in the context of broader trends in the political economy, state–family boundaries are shaped and reshaped by policy logic and various welfare reforms.

The second part of the book (Chapters 5 to 7) offers new insights and an in-depth understanding of people's views and practices based on the analysis of empirical data comparing the two societies in the East and the West. Chapter 5 presents an analysis of the ways in which families accumulate their housing assets. Comparative analysis is conducted in two layers between the two societies and also between the two generations. To complement their access to and experiences of home ownership, as well as their aspirations to become home owners, Chapter 6 focuses on people's views and practices regarding the disposal of their accumulated assets over the generations. Chapter 7 then returns to consider the key question of the book – whether there is a particular link between the provision of care and inheritance in the family, using the voices of older parents and their adult children in Britain and Japan. Given the complex and dynamic nature of family relations, the chapter examines whether people's attitudes match their practices and whether the views and expectations of the younger generation match the intention of their parents. These chapters also highlight the significant social change taking place in the two societies by looking at different views and attitudes of the younger generation. The book concludes with an analysis of the wider conceptual and policy issues regarding convergence and diversity in comparative studies.

2 The cultural practice of intergenerational reciprocity

Introduction

This chapter aims to explore in depth the reciprocal dimension of intergenerational relations, that is, the exchange of various types of support within the family over their life-course. It examines how patterns of family reciprocity have been shifting in the context of contemporary social change and how much such practices have cultural significance.

The role and structure of intergenerational relations have various dimensions. Finch (1989), for example, classified types of support in the family under five main headings: economic support (financial); accommodation (structural); personal care; practical support and child support; and emotional and moral support. Other scholars have added further categories, such as associational support (contact among family members) and consensual support (for example, sharing of opinions) (Lawton et al. 1994). Indeed, generations within families exchange not only material and financial resources but also instrumental and expressive support, including pride, love and status. The dynamics of support exchange are not only represented in such a wide range of 'exchangeable commodities' but can also be highlighted in a variety of ties that exist among different exchange partners within families. Although the main focus of this book is on the relationship between older parents and their adult children, support exchange can take place horizontally within the same generation between spouses and between siblings, and can also be extended vertically between grandparents and grandchildren, between in-laws and step-relations, or among more distant kin (Finch 1989). People's motivations and the level of obligations and commitments inevitably vary depending on their family role and other variables. While family members have always helped each other in most social contexts, in the context of contemporary social and demographic shifts, some recent studies started 'rediscovering' existing ties such as grandparents–grandchildren and also exploring relatively under-researched ties such as siblings (see, for example, Connidis 2005).

The focus of this book is largely on a particular exchange of housing assets and long-term care between older parents and their adult children within the family. Whether such an exchange is merely notional or in fact practised, theorizing wider exchange practice will help us understand the logic and capability of the exchange. Unlike the social contract mediated by the state, the micro-level 'generational contract' usually involves a direct exchange of goods and services within the family. However, there is a recognition that such micro-level interpersonal relations do not exist on their own but are influenced by and interact with macro-level social and economic structure and policies (Giddens 1991; Walker 1996). The role of the state or such interaction between structure and agency – 'individuals as actors exercise agency as they negotiate relationships within the constraints of social structure' (Connidis and McMullin 2002: 558) – will be dealt with extensively in Chapter 4, in particular with reference to the delivery of long-term care.

There is a clear cultural significance in exchange practice. Individuals in different societies are likely to possess wide variations in their conceptual understandings of intergenerational reciprocity; and how it should be or is implemented in their own lives. Given the different national contexts including culture, economy and public policy, this chapter examines how current social change in a globalizing world has brought new patterns to family relations, and how the particular relationships are negotiated within the family in different cultural contexts.

Social construction of exchange practice

Reciprocal arrangements are often a key to family support although reciprocity alone does not determine intergenerational relations. The norm of reciprocity often governs how individuals accept and provide support involving rights and responsibilities, and credits and debts (Akiyama *et al.* 1997). As Turner (1982) discusses, the focus here is firmly on exchange, as social interaction is often the process of exchange of materialistic or psychological rewards between actors on the basis of a norm of reciprocity.

Modern exchange theory, compared with the traditional functionalist approach, is more diverse and draws from various disciplinary sources such as economics, anthropology and psychology. For example, the idea of exchange as a source of or as a means to social solidarity has a long tradition in social anthropology (Craib 1992). While a classic economic theory on exchange defining such human actions as 'rationally seeking to maximize their material benefits from transactions or exchange with others in a free and competitive marketplace', contemporary theories are more realistic in irrationality in human action and macro-structural constraints (Turner 1982). As Turner (1982) summarizes, there are many assumptions and limitations in the economic theory as people are not always rational nor do they necessarily try to maximize profits by exchanging their utility. Such dynamics and unpredictability in human actions require profound analysis, in particular in the context of current social change. Moreover, people's transactions, whether in an economic marketplace or elsewhere, are not free from external regulation and constraint. Interpersonal relations within the family are always influenced by external factors such as laws, cultural traditions and availability/limitation of alternative resources as much as their own free choices and motivations.

The term *reciprocity* implies 'equal or comparable exchange' of various kinds of resources between individuals and groups (Akiyama *et al.* 1997). In this definition, exchangeable commodities do not need to be similar in kind as long as they are 'comparable' in conceptual values. Different commodities such as financial resources, actual goods, time and physical resources (practical and personal support), information and advice, or more symbolic expressive resources such as love and status could be exchanged between individual members. Exchanging different kinds of resources is quite common in some cultures, while in others exchange rules are defined as a more symmetrical pattern. For example, the exchange of different kinds of resources, such as money and affection, is quite common in Japanese families. However, such exchange rules differ among family and between non-family members because symmetrical reciprocity is the norm between non-family members (Akiyama *et al.* 1997). In the Japanese tradition, as Akiyama and her colleagues (1997) point out, expressive resources such as love, status and affection hold high currency and broad exchangeability. The high level of and commitment towards personal support provision by daughters-in-law have been driven by such moral values in exchange for their 'good' reputation. In contrast, the US family system prescribes 'symmetrical reciprocity' as an exchange rule. In this context, reciprocity – two-way exchange of goods and services – needs to be maintained in order to secure a good relationship even within intimate family relationships (ibid.). And resources tend to be exchanged in a short space of time in order not to be indebted to others for too long.

Exchange of family resources is usually based on voluntary action arising from mutual dependency. As Gouldner (1960) explains, if A and B have different resources and cannot survive without each other's resources, they would exchange their resources happily and willingly for mutual benefits. In this early theoretical work on reciprocity, however, Gouldner suggests that exchange can also create power relations if one has more resources and alternatives than the exchange partner, who may depend heavily on the other's resources. Also, exchange may not always be voluntary or even fair. For instance, if A is more powerful than B, A may force B to benefit with little or no reciprocity (ibid.). In reality, therefore, exchange may not necessarily take place as an equal relationship for mutual benefits to all the parties involved. In the family context, however, the redistribution mechanism often works and support tends to flow from the wealthier generation to the poorer one, which evens out inequalities. The demographic location of wealth over generations may thus determine the flow of resources in society, and an ageing process may create advantage or disadvantage in such power relations.

Exchange is only possible if two or more parties have some 'exchangeable' commodities, whether or not they are equal or comparable in value. Based on exchange theory, some existing data indicate that older people who can reciprocate resources are more satisfied with their life than those who only give, only receive or do not exchange (Kim and Kim 2003). However, Dowd (1975) argues that a disadvantage of old age in such exchange practice in the Anglo-Saxon context is that it tends to bring a necessary reduction in the possession of valued exchange commodities. In similar vein, modernization theory (Goode 1963; Cowgill and Holmes 1972) has also been provided as an explanation for the assumed decline in the status of older people in families and communities. In the process of urbanization and industrialization in society, an advantage which older people used to possess over skills and knowledge tends to be made redundant or become less relevant owing to the shift away from the land-based economy, family nuclearization, widespread education opportunity for youth and also ideological changes. According to Dowd (1975), such decline in valued resources results in a lessened ability of older people to interact successfully with younger people who have valued resources. However, it is true, to a certain extent, that such a status of older people may not paint a full picture of old age in some cultures as well as in highly advanced economies due to the following reasons.

First, 'exchangeability' has a cultural connotation. Expressive resources such as status, love and affection may hold high currency in some societies, which may not be reduced with age. Second, the demographic location of wealth over age cohorts shifts along with the modernization process and global economic change, which may alter the direction of support flow (Forrest 2005). In particular, older people in highly advanced economies can be more resourceful compared with their preceding cohorts and, in some cases, compared with successive cohorts. Today, while old age is not necessarily synonymous with decline in physical and/or financial health, it is true that older people, after reaching retirement age, tend to rely on a limited income such as a pension, which reinforces the social construction of old age. In Japan, however, the pattern of ownership of assets paints a different picture of old age. For instance, the cohort of older people appears to hold more than half of the nation's financial assets (and also tends to be outright owners of their property). Moreover, according to the US Census Bureau (1990), the group of older people aged 65 and over in North America in the 1990s was the wealthiest in history, since their median net worth and discretionary income was greater than those in the 'productive' age cohort between 45 and 54. In this scenario, older people may have more 'exchangeable' commodities including time (owing to not being engaged in paid work in the formal labour market) and financial and real estate assets, without many other competing responsibilities. In fact, due to the affluence brought to postwar families and the increased precariousness in the current labour market, parents supporting their adult children indefinitely, caricatured as 'parasite singles' (Yamada 1999), is not such an unusual practice.

Moreover, the timing or frequency of exchange may also vary. Intergenerational reciprocity can be considered over the life-course rather than the symmetric exchange pattern in a short period of time common in the US. Between generations, it is possible for the exchange to be one-way over the long term, if it is somehow reciprocated in the end. For example, parents may be a net 'provider' at a particular point in time due to a reasonable expectation that they may be a net 'receiver' at a later date (Akiyama *et al.* 1997). In Japan, as in many other East Asian cultures, the traditional norm based on Confucian teaching has been that the debt children feel towards their parents' sacrifices throughout their upbringing has to be reciprocated by caring for them in old age. For parents, therefore, having raised children earlier in life can serve as a credit for the receipt of support from them in their old age (Hashimoto and Kendig 1992). Chapters 5 to 7 look more closely at how such traditional norms have been shifted in ideology and practice.

Family reciprocity can be a continuous chain of obligations over generations rather than one particular parent–child relationship, especially in the Asian context, in which ancestor worship is often an integral part of family obligations. Research into Japanese migrant women growing older in British society, for example, suggests that geographical distance prevented those women providing care for their parents, and therefore their 'unfulfilled' feelings seem to reinforce their own attitudes towards not expecting support from their children (Izuhara and Shibata 2002). Having been away from their home society and crossing cultural boundaries for marriage, the majority of the female respondents were excused from their traditional duties as a daughter or daughter-in-law in terms of functional support for their parents(-in-law). In the situation of a British marriage, the role and responsibility of daughters-in-law is not structurally or functionally determined as they would have been had the women remained in Japan. Although this is an extreme case constrained by cross-national migration, the same exchange rules may be applicable to families geographically separated within the national boundary.

The question still remains whether the contemporary 'generational contract' is a notional concept rather than one implemented in practice, especially in the context of current global social change. Much existing research has indeed confirmed, for example, that reciprocity and affect are not necessary conditions for the provision of practical support, but instead feature significantly in the ideological construction of the caring relationship involving older people (George 1986; Qureshi and Walker 1989). Marshall *et al.* (1987) also argue that the nature of the caring relationship rests on a delicate balance between reciprocity, affection and duty. Such mixture and balance vary widely from society to society, and also over time in one society.

Filial piety as cultural practice

We cannot discuss intergenerational relations in the East Asian context without reference to the norm of 'filial piety' that traditionally defined *family responsibilities* and governed family practice in the region. The belief is often contrasted with the long-held values of individualism and privacy in the Anglo-Saxon nations that people have a duty not to be a burden to others. However, *familism* also co-exists with such individualism in those societies.

Filial piety is the moral norm of family ethics based on the Confucian doctrine and had a wider implication in the social system in East Asia. It was a foundation of the traditional patriarchal family system in many East Asian societies, including China, Korea and Japan, and has been (and to a certain extent still is) enthusiastically promoted by the state until the postwar constitutional and legal reforms started democratizing families in the region. Filial piety in East Asia, as described well by Hashimoto (2004: 182), is:

> ... at once a family practice, an ideology, and system of regulating power relations. As practiced in the family, filial piety defines a hierarchical relationship between generations, particularly that of the parent and the child. In this ordered space, filial piety prescribes the ideology of devotion by the grateful child to the parent, and also places debt and obligation at the heart of the discourse on parent–child relationships.

The family was patriarchal and the male head of the household held substantial power and authority over other members. Filial piety was indeed a gendered practice, emphasizing the father–son dyad. In Japan, for example, it was the duty of the eldest son, as the designated successor to the household, to perpetuate the family collectively, as family name, assets, social status of the family, and even occupation were usually inherited by the eldest son. He received a wife and continued an extended family household while daughters married out and second, third and subsequent sons formed a separate household. Debt and obligations cemented their generational bonds. Obedience was required in areas such as customary arranged marriage to favour family continuity over individual choice. Obligations were directed not only towards their parents but also vertically, including their ancestors over a few generations, by performing ancestor worship and continuing the family line.

Filial piety was indeed filial obligation and it was onerous. Janelli and Janelli (1997) argue that filial piety obligated children primarily to repay more than obey their parents, emphasizing the children's duty to provide care and respect for the parents' best interests. It was a gendered practice and the obligations were placed on the (eldest) son, but women (wives) performed the heavy end of care towards their parents-in-law through marriage. In this context, a strong emphasis was placed on the ties between the generations, especially between father and son, instead of conjugal ties (Sorensen and Kim 2004).

Although traditional family elements remained deeply embedded in the social structure of those societies, postwar socioeconomic and legal changes inevitably brought new ideology, functions and relations into the family. The constitutional and legal reforms in the postwar periods (the different stages in different societies in the region) dissolved the pre-war family system and the families in East Asia have been significantly democratized, which is evident in areas such as the increase in nuclear families, free marriage partnering, increased geographic mobility and inheritance practices. The discourse of 'new families' instead emphasizes equality, individual rights, freedom of choice and voluntary unions that denied many traditional aspects of patriarchal power and hierarchy (Hashimoto 2004). Contemporary filial piety may focus less on the aspects of obedience, although Hashimoto (2004) argues the contemporary version is an ongoing practice of surveillance and control over children. By examining current social issues such as social withdrawal, school absenteeism and teen prostitutions found in Japan, a hierarchy of power is indeed reproduced in everyday life, privileging older people over young people. Despite the abolition of the traditional family system, however, the notion of filial piety still remains strong in the caring dimension (supporting older parents) and ritual practices (funeral and ancestor worship), especially in rural communities.

People's perceptions, understandings and practices regarding filial piety have indeed been shifting. In contemporary Korea, for example, Kim and Kim (2003) found that reciprocal relations (older parents being able to exchange support in a timely manner with their children) are more valued than simply following the traditional norm of filial piety (parents' efforts to bring up children is repaid later in life). Contemporary filial piety derives from a variety of motivations. The 1992 survey conducted by Sŏng Kyut'ak in Korea found that the motivations of urban Koreans for upholding filial piety were a combination of love, respect, repayment of debts and a sense of responsibility or duty (Sorensen and Kim 2004). In fact, in this study love for parents scored higher than duty or sacrifice for the safety of parents. Given the increase in nuclear families, geographic mobility and women's social participation in East Asia today, managing and negotiating contemporary urban pressures to provide filial support to older parents may resemble the motivations and constraints which adult children are facing in the West.

There have been increasing regional variations in the practice of the filial support system. This can be explained by the nation's social, economic and institutional structures. For example, filial disengagement has been witnessed more strongly among the Japanese youth compared with their Korean counterparts. According to the survey conducted by the Prime Minister's Office in 1993, only 23 per cent of Japanese youth aged between 18 and 24 were 'positively' thinking of supporting their parents in their old age, in contrast to 67 per cent of their Korean counterparts. This could be explained by the fact that Japan, the most advanced nation in the region, went through profound social change, economic development and the establishment of social security much earlier than Korea. Compared with their Korean counterparts, current and subsequent older cohorts in Japan who worked in the economic growth period in the 1950s and 1960s are relatively well off, with their own savings and pensions. If filial piety is a means of pooling family resources together to survive in old age, available social security and welfare services may substantially undermine the motivations of the younger generation.

Moreover, even those societies that share the same cultural tradition can result in different filial practices owing to different political systems. Whyte (2004) found the paradox in the mid-1990s that the filial support system in urban Taiwan was more traditional than in the People's Republic of China, despite the fact that Taiwan was a much wealthier and more highly developed nation. This is partly because Taiwan's economy is dominated by small to medium-sized family-run enterprises, which helps to preserve the traditional system of co-residency, patrilineal kinship and pooling family resources over generations. In mainland China, in contrast, aspects of modernization were substantially accelerated by the socialist transformation of the mid-1950s when the family no longer became a 'production unit' (ibid.). As a result, intergenerational relations were much more a mutually beneficial exchange in China while the flow of support tended to be upward from adult children to their older parents in Taiwan (ibid.). Either way, the filial system of support appears to have survived well despite the rapid pace and contrasting trajectories of social change in these two Chinese societies.

In Thailand, a South East Asian society, where intergenerational relations are governed by the Buddhist religious norm (rather than Confucianism), there is a marked gendered difference in obligations. Ties binding parents and children, especially in less developed, rural communities, are governed by the asymmetrical reciprocity and deep-rooted morality of the Buddhist concept of obligations based on 'debts of merit' (Mills 1999). Here daughters shoulder a heavier burden than sons to provide material and financial support for their parental household. After marriage, daughters are the ones who form co-residency with their parents. In the religious context, parents and grandparents are responsible for transmitting the faith (the 'merit') and in return children carry out the beliefs and practices of their ancestors (Keyes 1983). In the Thai religious tradition, for example, children owe a debt to their parents and in particular to their mothers for giving birth to them. Sons can pay their debt back to their parents and sometimes grandparents by ordaining as monks for a limited period of time. Women, on the other hand, are barred from ordaining in the Theravada Buddhist tradition, and have to pay their debt only through respectful obedience to their authority and by contributions towards the physical and material well-being of the household, which is a much heavier burden that impacts greatly on their life.

This can explain the gendered pattern of migration from rural communities to urban centres such as Bangkok to seek employment to boost the household income. A strong dilemma is associated, however, with such practice since labour migration offers a means for daughters to fulfil their obligation to their parents partly by sending money back home. At the same time, absence of a daughter at home creates a heavier domestic workload for older mothers as well as the loss or delay of opportunity of receiving a son-in-law to supplement the household labour supply (Mills 1999). The cultural logic of intergenerational obligations has been increasingly challenged, however, since parental authority has been gradually eroded once the younger generations started being exposed to urban culture, ideology and materials. Overall, it is interesting to see whether and how Asian family obligations can survive in the current, more profound social change and globalization in the region.

Theories of intergenerational relations

The intergenerational *solidarity-conflict* and the intergenerational *ambivalence* models are two key conceptual and theoretical debates in the study of parent–child relationships in later life developed in the West. In more recent years, controversy has developed over these two competing paradigms since the two models tend to offer different conceptual lenses for understanding complex family relationships in societies undergoing social change (Lowenstein 2007). According to Lowenstein, these paradigms provide different ways to understand the interaction between macro-level structural forces and micro-level interpersonal relationships, and they derived originally from the different disciplinary backgrounds of social psychology and critical theory.

A large and growing body of literature suggests that the relationships between generations in families are the source of both solidarity and conflict (see, for example, Bengtson and Harootyan 1994; Walker 1996; Arber and Attias-Donfut 1999). Six dimensions of parent–child solidarity in two overarching categories were identified: affective-cognitive (affect, consensus and normative) and structural-behavioural (association, functional and family structure) (Bengtson and Schrader 1982; Roberts and Bengtson 1990). The advantages of the solidarity model lie in its focus on family cohesion as an important component of family relations, particularly for enhancing well-being in old age and longevity (Silverstein and Bengtson 1991, 1994). It appears to be designed to be sensitive to the multidimensional nature of family relations and also to ethnic and cultural diversity. The conflict model has then been developed to challenge the normative aspect of family solidarity (Marshall *et al.* 1993). The combination of these models is now generally referred to as the 'family solidarity-conflict' model, recognizing conflict as an integral part of understanding family relations (Silverstein *et al.* 1996).

In terms of such micro-level interpersonal relations, the solidarity-conflict model has been a popular framework for understanding family relations in later life for the past few decades; however, in more recent years there has been increased debate around such intergenerational relations, generating various types of *ambivalence* (Luescher and Pillemer 1998; Connidis and McMullin 2002). The shortcomings of the solidarity model, largely emphasizing positive aspects of family relations such as cohesion and consensus, have thus been scrutinized. Bengtson and Giarrusso (2002) argue that each aspect of the solidarity model may have multiple dimensions: for example, dependency and autonomy co-exist under 'functional solidarity'. However, the weakness of the model in not tying individual agency and social structure together has been questioned. Intergenerational ambivalence therefore offers another useful complementary (instead of competing) framework for understanding the more complex and dynamic nature of family relations by recognizing contradiction and paradox in parent–child relations in later life on both structural (sociological) and personal (psychological) levels.

Indeed, population ageing, globalization and current social change have increased the diversity and complexity of family lives and intergenerational ties. Owing to recent demographic shifts, the issue of generational equity as a source of conflict forms one of the major debates in welfare states (see, for example, Hills 1996; Becker 2000). Variations in cohort size (for example, the baby-boomers), changes in economic performance and shifts in policy and political direction over time are likely to generate inequalities among generations. There also exists a highly gendered conflict within the family as well as in society. In Japan, as in many Asian societies, for example, this partly originates from the traditional patriarchal family system where the roles and responsibilities of family members have been clearly defined. The role of women in the family, such as welfare producers, has indeed been socially and politically constructed, and cannot be explained without reference to macro-structural determination.

Family obligations could have become much more ambivalent with recent social change. For example, Giarrusso *et al.* (2005) argue that younger parents–adult child dyads will more likely be typified by conflict or ambivalence than their older counterparts owing to young people's competing commitments, work patterns and less stable family life. Many societies, including East Asia and transitional societies in Eastern Europe, are undergoing significant socioeconomic and demographic change in a comparatively short space of time. Such accelerated modernization processes and compressed development are likely to have a greater impact on the notion of intergenerational relations to create a generational gap in the area of social life. A more intensified generational gap being felt by current generations brings less consensus among generations due to shifts in financial wealth, policy measures (available alternative support) and the importance of pooling family resources. How these differential understandings are negotiated across the generations and in the context of development in social policy will be explored in the empirical Chapter 7.

Global ageing and patterns of intergenerational relations

Population ageing and globalization are part of significant contemporary social processes impacting on existing institutions, and they have also increased the diversity and complexity of family lives and intergenerational ties (Lowenstein and Bengtson 2003). Different patterns of intergenerational exchange emerge with changes taking place in generational patterns within families. Such demographic shift and institutional change in contemporary societies has created new pressure on families, yet at the same time have inevitably brought about the salience of family resources. The processes are likely to alter existing arrangements and to produce new, or redefine existing, patterns of intergenerational reciprocity. This section identifies a few converging as well as polarizing trends in intergenerational relations influenced by ageing processes globally.

First, a common policy concern in the global North for some years has been the impact of the ageing of populations caused by falling birth rates and prolonged life expectancy. This challenge will be accentuated over the next 10–20 years with the retirement of the postwar baby-boomers. A combination of increased longevity and low fertility means a shift from a vertical to a more horizontal structure, which Bengtson and Harootyan (1994) call the 'beanpole family' (see also Lowenstein 2005). Under the current demographic shift, there are more living generations ('intergenerational extension') but fewer members in each generation ('intragenerational contraction') (Bengtson et al. 1996). In this type of family, forming in many developed economies, adults have more ageing parents (generations above) than children. For Lowenstein (2005: 404), 'this process alters the length of time spent in specific family roles and leads to the emergence of adult children as the generational bridge between grandchildren and grandparents'. Also, more pressure will be placed on a smaller pool of younger resources for a prolonged period of time. This is coincided, however, with a privileged position of children whose share of the resources from above will be greater owing to having fewer siblings and wealthier parents/grandparents.

Second, the intensified speed of societal ageing brings a further complication to some societies. In this scenario, polarization is apparent between more established welfare states in the West and recently modernized ones in the East; and also between East and West in the European Union. In mature European societies, population ageing started much earlier (typically reaching the 'ageing society' of the 7 per cent level before the 20th century), resulting in the slower population growth of older people in the total population in the twenty-first century. In comparison, many Asian societies (and also South American), which have gone through a modernization process relatively recently, are experiencing an accelerated speed of societal ageing (see Figure 2.1). For example, Japan has doubled its rates of ageing from 7 per cent (UN-defined 'ageing society') to 14 per cent ('aged society') within 25 years (1970–95), while this doubling will happen in a far shorter time in Korea (19 years from 2000) and Brazil (predicted at 21 years from 2011). On the other hand, it took over 100 years in the already mature countries of Western Europe (for example, 115 years for France), and around 80 years in the developed Pacific Rim countries of North America, Australia and New Zealand, which are now reaching the 'aged society' level of 14 per cent (US Census Bureau 2001). Although the projection for China is more modest (27 years starting in 2000), the sociodemographic impact of the Single Child Policy since the early 1980s will bring about a distorted age distribution and will thus pose a serious policy challenge on the nation's dependency ratio, the future capacity and cost of care. Time usually gives society sufficient space for negotiating between generations and building intergenerational consensus over allocation and exchange of available resources. Rapidly ageing societies, on the other hand, tend to exhibit a much more profound generational gap regarding social norms, expectations and cultural practices. Such consensus and conflict models highlighting the East–West difference will be explored further in Chapter 7.

Third, global ageing has also brought the diversification of life-course and family forms (Lowenstein 2005). There previously existed the 'standard' life-course for men and women based on reproductive biology, the gendered social system and shorter longevity across societies. Today, however, such a conventional life-course has been altered and increasingly diversified. There is no longer such a thing as a 'normative' life-course found in many developed nations. For example, the British pattern of childbearing is now called 'twin peak',

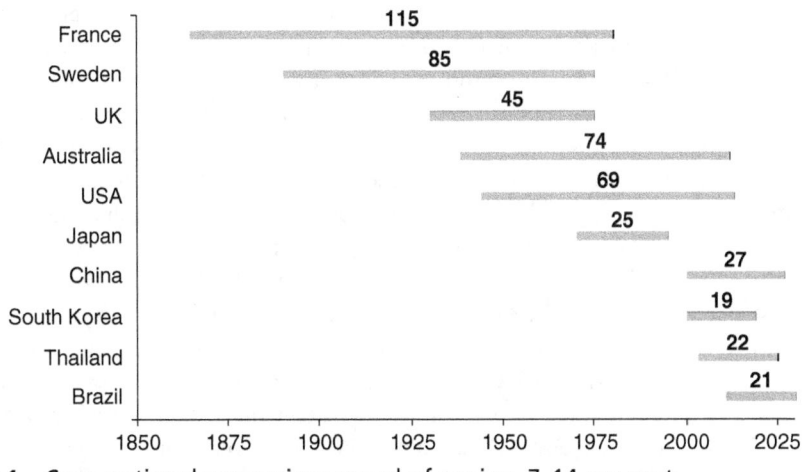

2.1 Cross-national comparison speed of ageing, 7–14 per cent.

with increasing teenage pregnancy (27 live births per 1,000 women aged 15–19 in 2004) and delayed motherhood (the peak age of childbearing is now in the early thirties). Bengtson *et al.* (1996) thus identified two distinct family types based on timing of fertility: age-condensed and age-gapped structures. In age-condensed families, the blurring of boundaries between generations occurs, especially when early fertility occurs across multiple generations, as among many Black families in North American society (Bengtson and Harootyan 1994). Age-gapped families, on the other hand, are increasingly common as many women choose to have their first baby later on in life. Such families may hinder the development of affective bonds and shared values across generations owing to the greater age gap (Lowenstein 2005). Therefore, age 'cohort' no longer determines their 'generational' role in the family, and thus age cohort may not become a useful measurable unit for comparative international analysis regarding family lives. Indeed, people in the same cohort may share less of their life-course-related experiences at a point in time.

Diversified family formation and structure present various national and regional specificities regarding family relations. One example increasingly witnessed in the West is the emergence of reconstituted families following divorce, separation, remarriage and re-partnering. These families are at risk of disruption and strain in intergenerational bonds (Ganong and Coleman 1998). In this scenario, the line of responsibilities may become unclear and need constant negotiation. This is likely to produce different patterns of family support between different ties over generations. The research in the US, for example, shows that adult children are obliged to help their divorced parents with minor tasks (but not physical caregiving) but tend to place priority on a parent over a step-parent, especially a stepmother. When choosing a step-parent over a parent for support, reciprocity was the most frequently mentioned rationale (for example, 'stepmother raised him') (Coleman *et al.* 2005).

In addition to following such a Western trend, a decline in co-residency is a more urgent issue that needs particular attention in relation to changing families in East Asia. Co-residency is one form of intergenerational solidarity (or conflict), notably strong in Japan as in many other Asian societies. However, postwar trends have shown a dramatic shift towards independent living, which has resulted in an increase in the number of elderly-only households in Japan. For example, approximately 70 per cent of people aged 65 and over lived with their adult children in the early 1980s while the rate dropped to 50 per cent by the end of the twentieth century. The detailed explanations for the shift and changing composition of co-residence families will be examined in the following chapter, but it should be noted here that the conventional 'generational contract' is now being transformed. Co-residency certainly provided a perfect structural context for exchanging family support, reinforcing responsibilities and sometimes making asset transfer easier between the generations. Such family nuclearization is thus likely to strain conventional family support motivations and practice. Structural changes do not always undermine the functional aspects of family relations, but such changes do require different policy responses (Izuhara 2000).

Another trend notable in developed societies is an increase in childless couples and individuals. Low fertility is one of the key driving forces leading to societal ageing, and differentiated patterns have been witnessed in different nations. In Japan, for example, the fertility rate dropped to a record low of 1.29 in 2004. The factors influencing this demographic shift are multifaceted and include both social and economic reasons, such as women's increased social participation and associated opportunity costs (Kawamoto 2001). The Japanese pattern of fertility is unlike that of Italy, where many couples tend to have only one child (Bettoni 2006), but more like the UK (some have two or three children while an increasing number are having none). In Japan, the fertility rates among married couples have been stable, at around 2.2 for the past three decades, which is enough to maintain the nation's population (Iwasawa 2001). There will be significant implications for those people who do not have children when they require informal resources for support in old age. Childless couples and individuals may place more importance on alternative relationships such as friends, siblings, nieces and nephews, and even formal services, to compensate for the lack of an intergenerational support system.

Social change and new patterns of reciprocity

With socioeconomic and demographic changes taking place in contemporary societies, relations between generations have moved away from conventional ideologies and practices. Such new patterns of intergenerational relations may be forming in the globalizing world owing partly to family changes, value shifts, new risks and precariousness associated with people's health and incomes, and also increasing mobility within and beyond national boundaries. The shifts may not be 'new' but rather a redefinition and re-emphasis of the already existing patterns of the family support system. This final section offers insights and an understanding of more contemporary exchange practices found in different regions and among different relational ties over generations.

We look first at the current debate around the role of grandparents in the intergenerational support context. Their role is defined in various ways, such as carer, replacement partner (confidante, guide and facilitator), replacement parent (listener, teacher and disciplinarian) and family anchor (transferring values, attitudes and history) (Harper *et al.* 2004). Some scholars (for example, Johnson 1983) argue, however, that given the voluntary nature of the role of grandparents, norms for their behaviour are less clear than for many other family roles. An ageing society means that people tend to occupy particular family roles for longer (Roberto and Stroes 1995) and having more generations in the family creates several generational ties (for example, great-grandparents and great-grandchildren).

Indeed, vertical linkages rather than horizontal ties in the family may have more emphasis in current demographic patterns (Harper 2004).

The opportunity for greater interaction across generations has increased owing to an increase in the number of living grandparents and an increase in healthy and wealthy people in old age (Uhlenberg 1980). Their involvement with grandchildren varies – some have more frequent contacts and exchange or provide support more often and extensively, while others have more infrequent, detached and ritualistic contact (Cherlin and Furstenberg 1986). As parent–child relationships, such strength and frequency of the ties are often determined by various influential factors such as age, gender, health, cultural norms, proximity and family lineage (Harper 2004). Grandmothers are said to be more involved with grandchildren, and maternal grandparents in particular are more likely to be called in for help when modern crises such as divorce and separation occur. Moreover, in some affluent societies, intergenerational transfer of family wealth may skip a generation and grandparents may be consciously leaving their assets to their grandchildren to provide timely support when the younger generation is in need and also to avoid double taxation when assets have been passed down twice over generations.

In many societies both in the East and the West, however, the role of grandparents in providing support such as childcare is not at all new. Grandparents have always helped their grandchildren (thus helping out their adult children both directly and indirectly). Comparing two different societies in the European Union, for example, regardless of the customary traditions and policy measures, Tobio (2004) found that grandparents provided significant amounts of childcare in particular for very young children in both France and Spain. Spain is a traditional Catholic society with strong familism, and mothers did not enter the labour market in large numbers until the end of the Franco regime; childcare services are not therefore well developed. In this sociopolitical context it is inevitable to witness the extensive support provided by grandparents. In France, where public daycare services are well established and two-thirds of couples are dual-earner households, both grandmothers and grandfathers (85 per cent and 75 per cent respectively) provide childcare support regularly, despite the fact that two-thirds of grandmothers are employed themselves. For Tobio (2004), grandparents as carers used to be sources for difficult times in the past such as war, death of a parent and divorce, but they now seem to be a common resource for normal times and for normal families. In the European context, care for small grandchildren by grandparents is increasing rather than decreasing. It is particularly true in the current economic climate where for women childrearing and paid work have become more complementary activities rather than alternative options (Land 2005). Land (2005), however, also argues that with an increasing number of grandparents forced to work into retirement, this valuable source of help may not be as readily available in the future.

The position of grandparents in the family may be expanding owing to an increase in divorce and complex reconstituted families in some societies. The contemporary dynamics of increased divorce, separation, remarriage and re-partnering indeed challenge conventional notions about family roles and obligations. Some scholars (for example, Dench *et al.* 1999) call it the rediscovery of grandparents and their importance in post-divorce families in the British context. Older parents are often called on during periods of crisis such as divorce and job loss, but such crisis appears to have become a much more common occurrence in contemporary societies. The UK literature suggests that there is an impact of divorce and remarriage on informal care provision for older parents.

Divorce/separation and forming step-relations appear to affect the ability of adult children to care for their older parents; and at the same time divorce and separation could reinforce a strong desire for older parents to remain independent (Dimmock *et al.* 2004). Again, such patterns are not necessarily consistent across families, because divorce and remarriage can bring family close together but can also create more distance between members, physically and emotionally. Despite the strong notion of fairness attached to dealing with children in British families, there is a certain dilemma for grandparents treating their natural and step-grandchildren in step-families (Dimmock *et al.* 2004). Grandparents may make a conscious effort to be fair for Christmas presents but when it comes to a disposal of real estate assets, they may use different criteria to decide who inherits what. As Dimmock *et al.* (2004) indicate, with an increase in divorce and separation (and such incidents occurring more than once for some people and later in one's life-course) it may be the case that the blood relationship between parents and children will become more reliable and important than those between spouses and partners. The evidence today suggests, however, that decisions over asset disposal seem to depend both on the traditional 'blood' relationships and the history of contact and maintenance, particularly between fathers and their children. Issues of property may well be separated from emotional partnership in those contexts.

Moreover, the issue regarding the missing middle generation is increasingly discussed in the developmental context and also in the US where the new risks in relation to health, crime and poverty are profound. Some grandparents act as replacement parents when their adult children (parents of grandchildren) are absent or incapable of looking after their own children owing to death, illness, imprisonment or labour migration (Burton 1992; Minkler and Roe 1996). According to Beltran (2000), the number of grandparent-headed households rose by more than 50 per cent over the past decade, partly owing to parental substance abuse, imprisonment, mental illness and HIV/AIDS. The role of grandparents can be much more extensive in such cases where, according to the US Census Bureau, 3.7 million children are being raised in their grandparents' homes.

In ageing sub-Saharan Africa, the endemic problem of HIV/AIDS has recently brought this issue into sharp focus. Grandparents have become even more important since many sub-Saharan nations have underdeveloped public resources to provide alternative support to families when they face the challenges of globalization. The development in sub-Saharan Africa is lagging behind, owing not only to endless civil wars, border conflicts and weak structures for democratic governance but also to HIV/AIDS, which is hitting the middle, 'productive' generation the hardest, devastating its human capital and introducing new trends in grandparents (Oduaran and Oduaran, forthcoming). Grandparents in many rural communities in developing societies also often fill a gap in family roles when their adult children, especially daughters, migrate to seek urban employment to boost their household income.

Finally, increasing geographic mobility in an era of globalization requires the redefinition of the informal support structure as well as the welfare regime structure of both host and home societies. Female migration in recent years has increasingly filled a gap in the health and social care labour sector of host (more developed) societies while at the same time vacating their capacity as a family resource in their home society. The movement of highly skilled labour is indeed an area that has recently attracted much debate in relation to migration, care and old age, especially in the European Union context (Ackers 2004; Harper 2005). Indeed such migration has an impact on families (dependent children, ageing parents and grandparents) left behind in the home society and poses a question as to how laws and welfare regimes can cope with the new patterns of care provision.

Conclusion

Filial obligations are socially constructed and this chapter highlighted the importance of culture in such a social construction of intergenerational exchange norms and practices within the family. Individuals in different societies tend to possess wide variations in their conceptual understanding of the intergenerational support mechanism and how it should be implemented in their own lives. With recent social change and demographic shifts, however, ideology and practice regarding intergenerational reciprocity appear to be increasingly divorced. Whereas a normative idea regarding traditional 'filial piety', for example, still remains in East Asia, the patterns of filial support, especially support from adult children to their ageing parents, may be converging to those practised in the West. Changing families and the development of public policy play a significant part in the changing patterns of intergenerational reciprocity.

Under increasing risks associated with family breakdown, a volatile economy, a precarious health and labour market globally, some generational ties such as between grandparents and grandchildren have been rediscovered and re-emphasized in some social contexts. In this scenario, the previously defined 'normative' life-course has been diversified and filial obligations based on such standard life-course have been significantly altered. Today family relations have become more dynamic and complex, requiring constant negotiation between the family members. The next chapter will explore the importance of family wealth in exchange of family resources. Housing usually represents the most valuable asset in the family and the chapter examines how such intergenerational transfer impacts on family relations.

3 Housing assets and intergenerational transfer in a global context

Introduction

In many capitalist economies throughout the postwar period, the growth of household assets, especially in the form of housing, meant increased prospects of inheritance. This chapter explores the role of home ownership in the accumulation of wealth over generations within the family, and how such intergenerational transfers impact on family relations in the current socioeconomic, demographic and institutional context.

The growth of the home ownership sector accompanying the accumulation of personal wealth has brought many advantages to households and society, and the first and foremost reason is economic. Home ownership can be associated with capital gain. Over the past three decades, in many home-owning societies patterns of housing price inflation indicate that housing has become a more important source of wealth accumulation. Studies examining the socioeconomic dimensions of home ownership and wealth accumulation were first introduced in Britain in the early 1980s as part of the rapid expansion of the sector fuelled by 'Right to Buy' – sales of council housing (Murie and Forrest 1980) and gained popularity throughout the decade. However, much of the debate was centred in Britain and other Anglo-Saxon contexts until the early 1990s (see, for example, Munro 1988; Forrest and Murie 1989, 1995; Thorns 1989, 1994; Saunders 1990; Hamnett *et al.* 1991).

As the following section suggests, residential property investment does not always accompany capital gain. The socioeconomic impacts of housing market bust and the consequent price fall have also recently been documented both in the East and the West in different periods, including the issue of negative equity in Britain (Forrest *et al.* 1999), the prolonged 'post-bubble' housing market slump in Japan (Forrest *et al.* 2003; Hirayama 2003a) and the impact of the Asian Financial Crisis in other East Asian societies (Forrest and Lee 2004).

Housing wealth is also important to people because the accumulation of such physical and financial capital helps individuals and families access and build other capital such as human and social capital (for example, investing in children's education, taking part in financial training) (Regan and Paxton 2001). Indeed, owning assets allows people access to security, independence and opportunities on top of the mere ownership of financial and material assets (Latham 2001). This process is likely to feed back into further accumulation of family wealth over generations. In many capitalist societies, accumulation of individual assets occurs largely through institutionalized mechanisms, primarily via home ownership and pension schemes that are clearly defined and heavily subsidized within public policy (Regan and Paxton 2001).

As many commentators point out, asset-building policies, including the post-war development of home ownership, can thus be seen as processes of social stratification (Forrest and Murie 1995; Paxton 2003). Such policies tend to benefit the middle class and create inequality and a wealth gap among families over generations. As Spilerman (2000) argues, social stratification derives not only from the individuals' rewards from labour market participation but also considerations of family asset holdings that have become increasingly relevant to general stratification analysis. Contemporary debates on asset-based approaches in public policy in the Western circle thus include the alleviation of such inequalities by trying to help those on low incomes accumulate assets through policy measures such as saving accounts, child trust funds and part ownership of social housing (Sherraden 1991; Regan and Paxton 2001; Chapman and Sinclair 2003). The former UK Prime Minister Margaret Thatcher's neoliberal approach in selling council housing in the UK could be interpreted as an asset-based welfare policy for low-income households through supported entry into the home ownership sector. The main intention may have been to extend the notion of a property-owning democracy (Saunders 1990) rather than privatization, but such economic and ideological objectives are likely to have co-existed in the policy measure.

The salience of home ownership has recently been revisited in the context of property-owning or asset-based welfare states (Groves *et al.* 2007). With the dispersal of neoliberal ideology and the intensified competitiveness of the globalized economy, there has been growing pressure to cut back on social expenditure and public subsidies. This has corresponded with the tendency of an increasing number of welfare states to promote individual ownership of residential properties. The idea is that the expansion of home ownership can be a deterrent to a high level of social expenditure, substituting for a decline in state welfare provision.

Home ownership has certain cultural dimensions. The meaning of home, accumulation processes and disposal of housing wealth may represent different cultural practices, which give cross-national, comparative studies particular resonance. In many industrial societies, where access to home ownership tends to be achieved largely by individuals' own means (through labour market participation) rather than inheritance, owning one's home may be associated with social status and rights of passage to becoming a 'full' adult (Dupuis and Thorns 1996;

Izuhara 2000), while for others, the ownership of land and home may be considered as 'family assets', largely granting continuity to the family. This may have originated from the idea that property is important for the structuring of family relationships. In an agrarian society, in particular, property has been a mode of production rather than consumption and controlling productive property (which provided the family with the means of obtaining an income) controls family members who are in a position to inherit (Allan 1982). This legacy of 'family continuity', or keeping the family line going, is indeed deeply rooted in some cultures. For example, family obligations and asset transfer have, in the past, been explicitly connected in the Japanese social and legal systems, and form part of the 'generational contract'. The effects of increased housing wealth within the family in contemporary societies are thus likely to create new bonds and new conflicts between generations around a shared interest in assets (Finch 2004).

This naturally leads to another key question – why is inheritance important or relevant in the context of family relations? The link between inheritance and kinship has been extensively studied in sociological discipline, primarily by Finch and her colleagues in Britain (Finch and Hayes 1994; Finch and Wallis 1994; Finch et al. 1996; Twigg and Grand 1998; Finch and Mason 2000; Izuhara 2004). In capitalist societies inheritance contributes to the accumulation of family wealth over generations. As Finch (1997) describes in the context of kinship, inheritance is only significant if there is 'something valuable' to pass on from one generation to another, which often means residential property rather than savings, since ownership of housing usually forms the largest share of an individual's or the household's assets (Saunders 1990). In this context, for example, given the concentration of land ownership before the industrial era in England, the creation of a new bourgeoisie whose wealth was based on industrial capital rather than land expanded the number of families for whom inheritance became a relevant issue (Finch 1997). Even in Britain, which has one of the highest levels of home ownership in the world, however, the expansion of the sector has only been seen in the postwar period and over the past two decades in particular. Thus, inheritance has only recently become a concern to the majority of the population, when the first generation of home owners started entering old age en masse. Indeed, inheritance not only contributes to family solidarity but also causes conflict. This point is highly relevant to this book – the reciprocal dimension of intergenerational relations exchanging support and housing assets within the family, as discussed in the previous chapter.

Finally, contemporary debates around the world regarding housing wealth accumulation have shifted and now include the importance of its strategic potential as an income source in later life. The issue of using housing assets to fund the cost of care in old age is, however, new to social policy (Finch 2004). This is debated partly in the context of the global economy, labour market restructuring and family and demographic change, which combine to produce uncertainty in future pension funds and return from investments. And assets of older people tend to be tied largely to their housing. To secure and supplement their income and to meet the changing needs of housing in old age, there are various options that older people could deploy to release part or all of their equity tied to housing (Izuhara 2007). Another policy debate that is central to this book is how the state perceives and treats individuals' housing assets when considering the cost of care (see Chapter 6 for a fuller discussion).

This chapter (and the following chapter) aim to set a wider theoretical and conceptual framework considering trends and developments across the OECD (Organization for Economic Co-operation and Development) societies. The comparison also goes beyond Western examples by including East Asian societies, so that the cases of Japan and Britain stand out. It will also aid analysis of data on micro-level interpersonal relations drawn from the empirical research for later chapters.

Processes and outcomes of housing wealth accumulation

As the level of home ownership is often used as an indicator to assess the nation's inheritance potential (Thorns 1994; Forrest and Murie 1995), this section attempts to map out the growth of this sector across various home-owning societies, and to highlight some critical issues associated with the accumulation processes and outcomes. Mapping out the current state of the sector will also shed light on where Britain and Japan are located globally. It is worthwhile noting, however, that residential property transfer often forms the largest but only part of an overall asset transfer within families. Identifying the level of home ownership in society can thus help us understand the potential number of households that are likely to inherit housing wealth. But what it does not tell us is the patterns of such bequests and their contribution to intergenerational transfer, and wealth transferred through other modes of intergenerational transfer including *inter vivo* transfers during people's life-course. According to Gale and Scholz (1994), for example, 20 per cent of US personal wealth passed on to the next generation was as gifts during parents' life-time.

In the 2003 Housing and Land Survey of Japan, intergenerational transfer only appeared under the question 'the means by which households obtain their dwellings' (9 per cent of owner-occupied properties were obtained through inheritance or gift) (Management and Coordination Agency, 2003). However, the survey captured only those who occupied their 'inherited' property but was blind to other intergenerational transfers that may have fed back to housing wealth. Moreover, some parents help or invest in their children throughout their life-course but not all such *inter vivo* transfer is officially recorded. Passing money to children or directly to grandchildren early is one of the most efficient ways of avoiding inheritance tax in some societies, and in the UK there is a £3,000 annual gift exemption, for example. In Japan, allowance for such *inter vivo* transfer is even more generous in the current economy. Tax-free cash allowances were raised from ¥1.1 million (£5,500: £1=¥200) annually to a total allowance of ¥25 million (£125,000) in 2003. Additional allowances for house purchase and improvements are also generous in Japan, reflecting the current government's intention to stimulate the economy by providing families with an incentive to redistribute rather than save their wealth. This system also helps family wealth find its route to the owner-occupied sector. Co-residence is another way to redistribute resources across generations and tends to contribute to a reduction in inequalities among generations (Attias-Donfut and Wolff 2000). But again, owing to the informal nature of household sharing, it can be difficult to trace statistically how much wealth/resources is transferred between the generations.

Doling (1997) indicates in his comparative analysis of selected advanced industrialized societies that home ownership has become statistically more dominant in these societies during the postwar period. Among the selected 11 societies, predominantly in Western Europe, the average size of the sector grew from approximately 48 per cent in 1960 to 58 per cent by 1990. But there are some distinct variations. First, the rate of increase varied considerably from country to country with the UK scoring the highest (almost 40 percentage points increase between 1945 and 1990), and only a 7 percentage points increase in the US during the same period. One explanation for such a dramatic increase in the UK is the Right-to-Buy policy for social tenants that has played a crucial role in boosting the level of home ownership since the 1980s. Such privatization initiatives, however, only work in societies that have had a period of intense public housing development. Second, some societies became home-owning societies considerably early. In Australia, 50 per cent of its households were home owners in the 1890s (Butlin 1976, cited in Mullins 2000: 684) – a rate achieved by Britain in the 1970s. Therefore, a large proportion of Australian families would have accumulated housing wealth over the twentieth century and already been in a far better position to pass their equity on to the next generations (Mullins 2000). Third, in some societies, the increase in owner-occupation has plateaued out, and Australia, the US, Denmark and Japan, although at different levels, are some examples. Finally, the current levels of home ownership vary considerably across the societies, ranging from 30 per cent to over 80 per cent. As Doling (1997: 160) states, 'even within small groups of countries with similar cultures, histories and economic development, such as the Nordic countries or those of the Iberian Peninsula, there are wide differences'. For example, in 1990 the rates were 58 per cent in Portugal but 76 per cent in Spain. Such differences can be explained by the differences in their institutional context, and this point will be returned to later.

Within the European Union (EU), variations are rather striking between new and original Member States. The home ownership rate is in general extremely high among the new Member States – over 80 per cent (except for the Czech

Republic and Latvia, which is associated mainly with mass privatization policies). Among the former EU15, on the other hand, the rate is above 70 per cent only in Ireland, Italy, Luxemburg and Spain, and none of the other EU15 countries exceed the 80 per cent line (Eurofound: Communiqué, Issue 5, 2004). Interestingly, as in East Asia, the level of home ownership varies inversely with GNP (gross national product) per capita since the poorer countries tend to have higher levels of home ownership.

In East Asia, the rise of home ownership is a relatively recent phenomenon except for Japan and Singapore, and has been closely associated with the rapid economic growth of 'tiger economies' since the 1970s. Unlike the Western experiences where the market dominates, the productivist notion of the developmental state signifies that the promotion of home ownership is part of social and economic policy. Thus, home ownership is more closely associated with direct public intervention than the private market (Lee *et al.* 2003). The levels of home ownership are generally high in the region (86 per cent in Singapore in 1999; 85 per cent in Taiwan in 1997 and 75 per cent in Korea in 1995) apart from Hong Kong, in which half of the population continues to rent public housing. Japan is another exception in the region in terms of the level and growth rates – over the past few decades, the Japanese rates have been rather stable, at around 60 per cent. The stagnated growth in recent years can be explained by the late entry of younger people to the sector and also the increase in single-person households (who tend to rent). It is true that, as in Europe, the wealthiest societies/city states have the lowest levels of home ownership (60 per cent in Japan and 52 per cent in Hong Kong) and the highest housing price to income ratio (11.6 and 7.4, respectively) (Lee *et al.* 2003: 22). Forms of ownership vary substantially in the region; for example, the majority of home owners in Singapore own public housing while such arrangements virtually do not exist in Japan. House price volatility has been a shared experience among East Asian home owners due largely to the economic downturn they have experienced since the 1990s, notably the burst of the bubble economy in Japan and the Asian Financial Crisis in the tiger economies.

There are some institutional mechanisms to encourage or discourage individuals and households to accumulate their housing wealth. In other words, the way in which people accumulate their wealth is embedded in the nation's socio-economic and institutional context. Over the first half of the postwar period, common global features that boosted the development of the home ownership sector were rapid economic growth, rising real incomes, growing job security and stable families. And the development of financial markets accompanying the greater availability of housing loans is another reason. Country-specific variations are thus likely to come from their institutional and policy context. In some societies, the growth of the sector has been largely assisted institutionally through a combination of housing provisions, assistance (for example, low interest loans), incentives (such as tax breaks) as well as the lack of alternatives.

Housing assets and intergenerational transfer

In Japan, the relatively high level of home ownership has been achieved through a combination of state-subsidized loans, occupational welfare (company housing acted as a stepping stone to home ownership) and the limited provision of public housing. The Government Housing Loan Corporation (GHLC) (abolished in 2007) had the major responsibility of providing heavily-subsidized low-interest loans for home ownership to most middle-class households (Hirayama 2003b). Such subsidies are also common in other societies such as Singapore, where, over the past two decades, forced savings in the Central Provident Fund could only be withdrawn early for housing mortgage financing. And the Housing and Development Board in Singapore provides loans to purchasers of its leasehold flats below the prime rate of a commercial bank (Phang 2001: 449). In the current economic climate, however, many generous (or even less generous) state assistance towards home ownership has ceased to be available – for example, the abolition of the GHLC in Japan and the abolition of MIRAS (mortgage interest tax relief) in 2000 in Britain, which may represent the end of the governments' support towards home owners.

Differentiated outcomes: asset accumulation and inheritance

The accumulation processes and outcomes are not homogeneous across societies. And even within a society, such processes and outcomes may vary significantly according to gender, age, occupation, income levels and family backgrounds. Factors such as timing of purchase, housing type, age of the stock and location also dictate the level of assets households own, which contribute to the accumulation of family wealth over generations: there are also marked cross-national differences.

Housing market trajectories: until the crisis in sub-prime mortgages hit the US property market in 2007, there has recently been a worldwide boom in housing prices. Two common factors underpinning this property boom are the historically low interest rate and the slump in the stock market as an alternative investment. According to *The Economist* (18 June 2005: 73), in 2004/5 housing prices continued to rise by more than 10 per cent in many countries such as South Africa, the US, Hong Kong and New Zealand. France (15 per cent) and Spain (15.5 per cent) had faster house price inflation than the US in the first quarter of 2005. By contrast, some housing markets were stagnant. The British and Australian markets have cooled significantly, while prices fell in Germany and Japan in the same period.

Britain and Japan have presented differentiated housing market trajectories with marked regional variations since the 1990s (Hirayama *et al.* 2003). In the 'golden age' of home ownership until the first half of the 1980s in England and the late 1980s in Japan, housing prices consistently rose and home ownership was

Housing assets and intergenerational transfer

accompanied by capital gain (Saunders 1990). Under such circumstances, people aspired to become home owners and expected to accumulate assets through housing price inflation. Over the past two decades, however, housing markets have become more volatile in both societies. Booms and slumps have been experienced most sharply in London and the South East of England, and in major Japanese cities such as Tokyo and Osaka. As Hamnett (1999) points out, there have been big winners in terms of capital gain, depending on when a house was bought, its location, how long it was owned and the social characteristics of the owners; however, there have also been big losers. The recovery of the British economy from the last recession has continued to grow since the start of the new millennium. During the 1990s housing prices rose 69 per cent on average from £63,173 in 1990 to £106,998 in 2000, with some distinct regional differences (DETR 2001). The Japanese housing market, by contrast, has remained depressed for over a decade, owing to the prolonged post-bubble recession (see Figure 3.1).

The nature of the housing stock: size and type of properties, age of the stock and ownership status all dictate the level of housing assets. In terms of housing stock, the appreciation of assets goes to different types of stock in different national contexts. For example, in Britain older 'character' properties such as Victorian and Georgian houses are more highly appreciated while the opposite is the case in Japan. Owing to the 'scrap and build' approach driven by the construction industry, combined with other factors such as the type of materials used, the expected lifespan of Japanese houses is much shorter. It is thus not too much of

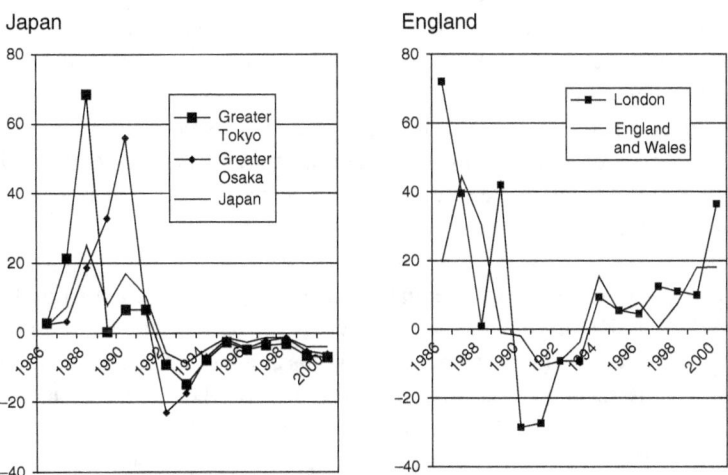

3.1 Residential land prices in Japan and England (1986–2000).
Source: Ministry of Construction (Japan); Inland Revenue Valuation Office (England).

33

an exaggeration to say that by the time the younger generation inherit residential property from their parents, it may 'require' rebuilding.

Ownership status (freehold, leasehold or shared): ownership status also influences the values of properties as tradable commodities. Some societies have seen more condominiums (and also in some East Asian city states where flats and condominiums are a dominant form of home ownership); collective ownership is one of several issues regarding this form of ownership. In Japan, this type of owner-occupation has gained in popularity, especially among young families in urban areas, given the scarcity and extremely high cost of urban land. Owing to rapid depreciation in value, however, in particular suburban condominiums that were oversupplied in the 'bubble economy' period, much of this type of property now functions primarily for use rather than as an investment. Second-hand condominiums had experienced a steeper fall in value and have continued to do so in the context of an apparent bottoming out of prices in other sectors (Forrest *et al.* 2003).

When successive cohorts of condominium owners who failed to trade up to a single-family home start to enter old age, the implications of condominium ownership on property inheritance for successive generations will be much more profound. Condominium owners who face the necessity of redevelopment in their old age may, reluctantly, need to move back to the rental sector, given their limited income and considerably reduced housing equity. The future of shared ownership thus poses many questions, especially since the reform of the Sectional Ownership Act in 2002 in Japan. Under the revised Act, the redevelopment of condominiums can proceed with the agreement of the majority (80 per cent of the owners) instead of full consensus, which excludes the minority's views from the decision. When the cost of demolition is subtracted from an already small share of individual equity from a plot of land, it is almost impossible to buy back a unit in a newly built building on the same site. In the UK, sales of council flats (rather than houses) since the late 1980s have also created new problems associated with leaseholds regarding management and administration (Forrest *et al.* 1995). The issue was around how to create collective agreement on maintenance and repairs of shared facilities and spaces in a building.

Between genders: gender differentiation of housing asset accumulation tends to be overlooked in much mainstream research, particularly because housing-related surveys are often conducted among households as a unit. In comparison, financial assets such as savings, investment and pensions, can be more personalized and in some societies may reflect women's financial status as individuals more accurately. This is especially the case in societies such as Japan where a 'joint account' or 'joint mortgage' are not common. A limited number of studies are currently available to reveal a complete and comparative picture of women's housing asset holding globally. But in general, women's perceived lack of access to home ownership can be associated with their labour market position and participation, income and the discourse around the family and the male-breadwinner family model (Hirayama and Izuhara *forthcoming* 2008). This can be linked to institutional discrimination and inadequate resources, as well as the social norms and gender role definitions, although such socioeconomic and cultural conditions often peculiar to each society tend to produce different outcomes. The high level of home ownership found among older women or female-headed households can be a result of inheriting housing property from their late husband.

Comparing Britain and Japan provides an interesting case study. According to the survey conducted cross-nationally (see, for example, Izuhara and Kennett 2006), despite some common disadvantages women are facing in both societies, significant differences have been observed in terms of their housing asset holding. While tenure structures are similar, women's access to the asset accumulation in land and property ownership varied significantly between Britain and Japan. Marital status is an important indicator of women's housing circumstances: in Japan, the majority of married female respondents were owner-occupiers but the property was most likely to be in the name of the husband (73 per cent of owner-occupier respondents). In Britain, on the other hand, it was common to have joint ownership and a joint mortgage among married and co-habiting couples (76 per cent of those who were buying their property through a mortgage and 21 per cent of female borrowers were the sole mortgagor) (Izuhara and Kennett 2006). For British women, higher rates of divorce, debates around gender equality, changing cultural norms as well as rising house prices may have encouraged women to 'stake their claim' in relation to housing.

Gender difference may be apparent in inheritance practices across societies, which also helps accentuate differentiated patterns of wealth accumulation between men and women over generations. Family wealth going disproportionately to the next generations of men due to the power of patriarchy has been common practice in many Asian societies, as indicated in Chapter 2 (see, for example, Tokyo Women's Foundation 1997). However, such practice has also been found in the Western context, in particular passing on *market* rather than *domestic* property (Delphy and Leonard 1986; Pratt and Hanson 1991). Other studies in the West are more likely to claim that such intergenerational transfer of residential property tends to be made equally to the next generation regardless of gender (Finch and Hayes 1994; Thorns 1995; Mullins 2000). The line of descent in the intergenerational transmission of family wealth in relation to inheritance laws and customary practices will be examined in more detail later in this chapter.

Among generations: although this book is less preoccupied by the 'demographic location' of housing wealth, since the main focus of wealth transmission is from the older to the younger generation, the so-called 'cohort effect' of accumulation patterns can be peculiar to each society. According to Forrest (2005), such cohort effect has particular resonance in an East Asian context largely because significant differences in the region's economic conditions and institutional contexts are likely to have produced varying entry conditions for different generations. In Japan, like the mature home-owning democracies of the West, the weight of housing assets is likely to be located at the older end of current generations with potential flows of material and financial assistance from old to young. Their assisted entry to home ownership was largely through postwar economic growth, occupational welfare and public loans by the GHLC. In Hong Kong, in contrast, Forrest (2005) claims the baby-boomer generation is the one that has been most advantaged in terms of policy regimes and economic cycles. Their economic fortune is often contrasted with the previous generation of refugees from Mainland China and the successive cohorts who face a more competitive job market, a more uncertain housing market and less stable public assistance in gaining entry to home ownership.

Older people are in general more likely to be outright owners of their housing because of the entry and duration of their owner-occupation. This is an important factor since the households' overall assets consist of both positive and negative assets, and the younger generation can inherit family wealth without debts. There are, however, some older people who enter the home ownership sector relatively late and still have mortgaged housing equity. For example, as a result of purchases under the Right-to-Buy scheme in Britain, the proportion of older people not owning their home outright increased during the 1990s (Leather 1999). Becoming a home owner in later life is also possible in Japan, since financial institutions do not usually discriminate against borrowers by age, and intergenerational loan inheritance schemes have become more widely available (Izuhara 2002). Such schemes help maintain the customary tradition of co-residency in Japan by enabling two generations to purchase a house with a longer repayment period. Intergenerational mortgages are also available in Western societies, such as Switzerland and Ireland. More recently, a no fixed term intergenerational mortgage has been introduced in the British financial market (*The Guardian Unlimited*, 23 August 2006). Under the interest-only scheme, children inherit a home as well as debt and it would help cut the amount paid in inheritance tax. This may be an advantage to people on low incomes to enter home ownership.

Impacts of social policy: regarding housing for older people, the international policy trend is 'ageing in place', which encourages people to remain in their own home for as long as possible with formal and informal service delivery and necessary housing adaptation. The debate around housing assets in old age indeed links to the issue of long-term care. Such 'ageing in place' policies prima facia enable older home owners to retain their assets invested in housing, but how the state perceives and treats individuals' asset accumulation manifests fundamental differences across societies (Izuhara 2005). For example, in England (Scotland has a different policy), the housing assets of individuals have increasingly been seen by policy makers as a source of income in old age. Under the current system, people with capital or savings of more than £18,500 will be expected to meet the full cost of their long-term care fees. The necessity to sell a home in order to move into residential or nursing homes has been the source of considerable public and popular concern and attracted extensive debates around the issue (see, for example, Deeming and Keen 2003; Wanless 2006). Japan, on the other hand, has taken a contrasting strategy to fund increasing care needs in an ageing society and individuals' income and assets are irrelevant in the assessment process to receive public services for long-term care. This point will be analysed in relation to long-term care policy in the following chapter.

Demographic change and inheritance

Demographic change influences how family wealth, in the form of housing, trickles down and benefits future generations. In many advanced economies over the past 25 years, changing families have indeed started to alter preconceived ideas regarding property transfer. Such change has been happening in the context of increased housing equity among current and successive older generations. While some characteristics are shared among many Eastern and Western societies, others are more specific to a particular national context. For example, inheritance practices are highly relevant to and influenced by population ageing, which is one of the most significant demographic changes currently taking place in the global North.

Britain provides an interesting case study having experienced both a rapid increase in home ownership and a significant change in family structure and formation in the postwar period. While greater housing equity is now available within households to pass on, the profile of its potential beneficiaries looks rather different. As Williams (2003: 165–7) points out, in the first half of the postwar period, when the home ownership sector grew, an individual's life-course was relatively predictable – rising real income, job security and a stable family. However, 'families, jobs and incomes are now less stable, life expectancy has increased considerably and with the continued growth of home ownership the likelihood of being an owner-occupier and inheriting a property or the proceeds of a property are now considerably greater' (Williams 2003: 166). Issues related to housing equity transfer in the context of demographic change are now discussed.

Increased longevity: societal ageing is largely a result of a combination of increased longevity and a decline in fertility. Life expectancy in advanced nations now exceeds well over 80 for men and for women. The first concern is therefore the passing end (testators) of the spectrum influenced by increased longevity. In terms of intergenerational asset transfer, increased longevity means that two adult generations in the family exist, often as independent households, for a long period of time, which delays the transfer of wealth over generations; and also between spouses (see Izuhara 2005). Further delays are occurring in contemporary Japan, where the transfer of assets no longer skips widowed wives as it used to under the pre-war patriarchal succession system. The timing of inheritance conditions the advantage it offers to the younger generation, for example, entering home ownership, trading up or improving property, paying off a mortgage, or supplementing resources for childrearing. Given the prolonged longevity of people, however, a growing proportion of adult children are likely to have an established household with a mortgaged or unmortgaged property by the time they inherit assets on the death of their parents.

Existing studies in the West (see, for example, Hamnett *et al.* 1991; for UK: Holmans and Frosztega 1994; for Canada: Thorns 1994; for Australia: O'Dwyer 1999) revealed in fact that most beneficiaries (excluding spouses) tended to inherit when they were already outright home owners. What are likely scenarios if home owners solely or jointly inherit their parental property? If property inheritance is made to multiple beneficiaries, which is common in many Western societies, options are to liquidize it by selling it, to transfer ownership to particular beneficiaries, to sell it to another beneficiary or to become a landlord. The Australia case suggests that housing inheritance in this context may offer an opportune new source of private rental housing, although such amateur landlords have different expectations and experiences in the private rented sector (O'Dwyer 1999). Another likely consequence reported by the International Longevity Centre (2003) in Britain is inherited wealth skipping a generation and going straight from the estates of older people to their grandchildren. According to a study conducted by Finch and her colleagues, however, there was little evidence to support this claim in the UK context recently – only 12 per cent of wills contained bequests to grandchildren and the typical grandchild bequest was a small cash gift (Finch 2004). In less than 2 per cent of cases, a grandchild received a share of the total or residuary estate.

Given the accidental nature of bequests under the life-course model, wealth tends to flow from the old-olds to the young-olds (adult children), and thus remains stubbornly in the hands of the older age groups and is subsequently underused or unused (Arakawa 2003a,b). The survey conducted by the Daiichi Life Insurance Research Centre in 2002 in Japan highlighted that inheritance received later in people's life-courses had little impact on their consumption or investment patterns. For example, for 81 per cent of those who had inherited assets, there was no significant effect in their consumption and investment pattern after inheritance. Rowlingson and McKay's survey (2005) also revealed that inheritance was unlikely to raise the living standard of people with poor incomes in later life, and those who received substantial amounts were already relatively affluent and thus it had little impact on their life. However, when inheritance starts skipping a generation and benefiting more substantially younger people with their first entry or subsequent upgrade of their housing, the future impact of inheritance is likely to be greater. The importance of breaking the current cycle of such underused or unused assets among older people has become a major debate in society by encouraging them to spend more or transfer wealth earlier to younger generations (Arakawa 2003b; Izuhara 2007). The impact of inheritance on younger generations needs further exploration given current socioeconomic changes. Furthermore, increased longevity adds to uncertainty about how much wealth older people will consume before their death. This point will be discussed in more detail in Chapter 4.

Low fertility and high mobility: low fertility is the other side of the same coin of societal ageing, and also has a significant implication on property inheritance. Having fewer children in a family means in theory that children are likely to inherit a larger share of assets if every child inherits equally. In low-fertility societies, evolutionary theories of the family suggest that 'parents act to maximize their investments in their children' (Lye 1996: 81). Evolutionary theory also hypothesizes that parents will provide the most resources to adult children who have or who are more likely to have children. In this scenario, an unequal distribution of wealth may occur, even in a society with a strict determination of fairness over generations.

There is a certain cultural dimension to such practices. The issue of choice of beneficiaries as customary practices or legal rights of children illuminates one of the most striking differences across societies, since different laws and practices can produce differentiated patterns of intergenerational asset transfer. Moreover, as Forrest and Murie (1989) point out, the nature of property inheritance in relation to demography may have a strong class dimension. For example, in Britain working-class home owners tend to have smaller housing capital yet a greater family size and thus a much reduced propensity for such forms of inheritance among children compared with their middle-class counterparts.

Geographic proximity between the generations also makes a difference to the type and amount of bequests that children receive. In Japan, those who have been less geographically mobile are more likely to inherit among siblings (Tokyo Women's Foundation 1997). However, there may be a contrasting outcome in relation to class and geographic mobility in the British social context. Less proximity is probably more common among middle-class families but they may be more likely to inherit a larger amount including residential property.

Changing families and households: changes have been observed in the formation, structure and function of the family. First, a notable international trend has been the move towards smaller households. An expansion of single-person households (including among older people) has been witnessed for the past three decades. In particular, in relation to intergenerational transfer, a decline of intergenerational households (co-residency) is a notable trend in the East Asian context.

This type of household is still fairly common in Japan as in some other Asian societies compared with their Western counterparts. According to the 2001 round of the International Survey of Lifestyles and Attitudes of the Elderly, the proportion of older people aged 60 and over living in three-generation households was only 2 per cent in the US, 1 per cent in Germany and virtually 0 per cent in Sweden, but 22 per cent in Japan and 26 per cent in Korea (Cabinet Office 2002, cited in Ogawa *et al.* 2006). However, the Japanese rates have clearly been in decline for the past three decades. The Japanese government has viewed this pattern of household structure as a unique asset, and tried to preserve it through various policy measures such as tax breaks for those who live intergenerationally, generous cash allowances for housing-related *inter vivo* transfer and a mortgage inheritance system. Co-residency certainly provides the perfect structural context for exchanging family resources, reinforcing responsibilities and making asset transfer easier between generations. Further accentuation of family nuclearization has meant that the traditional practice of living with the eldest son (and his family) has been under threat, which has enormous implications for the traditional system of one-son succession.

Another international trend notable here is a decline in marriage and an increase in family breakdown. As discussed in Chapter 2, more people are thus living out of the 'conventional' family structure owing to increasing rates of divorce, remarriage and single-person households. Complex families such as stepfamilies are likely to complicate inheritance practices partly because reconstituted family members may hold different financial backgrounds, expectations, responsibilities (towards both divorced and stepfamily members) and legal rights (Burgoyne and Morison 1997). The shift away from the normative family model also means that some women need to seek alternative means (to marriage) to secure their housing. Such a scenario immediately raises an issue regarding women's access to housing and accumulation of wealth. Overall, changing demography contributes to a diversification of people's attitudes and practices towards asset transfer.

Socio-legal context: Family relations and inheritance laws

Who are the likely beneficiaries among family members? What determines their share of bequests? Are family obligations reflected in inheritance laws? The answers to these questions illuminate some of the most significant cross-national differences in inheritance laws and practices. For comparative analysis we use Japan and Britain primarily as examples.

In Japan, constitutional reform after the Second World War brought democracy to the family. Until the traditional form of the family system was abolished by the new civil code, eldest son succession was the norm reflecting the patrilineal stem-family system. Under the new Family and Inheritance Law in 1948, rules of inheritance were redefined according to the new definition of family. The new inheritance codes are clearly based on the principle that certain family members, especially spouses and children, have the right to an equitable share of family assets. Spouses are legally entitled to receive half of the deceased's assets, while the other half is divided equally among the children. A variation to include extended family members only occurs if such immediate members are absent. For example, if a child has died their share can be passed onto their children (grandchildren to the deceased). If a deceased child was married but without children, but the deceased had siblings or surviving parents, the siblings or surviving parents would be entitled to a (relatively small) share of the assets. Since the traditional family system used to determine the ways in which family wealth was passed on, people have not been accustomed to making wills. Even if assets were left in a will, the immediate family members could legally claim half of the total assets. Thus it is unlikely that the entire wealth would pass to a third party. In reality, however, many informal arrangements still supersede such legal definitions and passing on housing assets intact to one particular child is not uncommon.

The disadvantaged position of daughters (especially married daughters) and non-successor sons is distinctively 'Asian' (although in many Western societies women have historically also been disadvantaged in asset accumulation and inheritance). A common practice found in many Asian societies such as India is that a dowry (*inter vivo*) is given to daughters at marriage while inheritance is passed on to son(s). However, changes have also started taking place in recent years in other parts of East Asia, although much later than in Japan. In South Korea, for example, a highly significant move came in the 1990 legal reform which allows daughters to inherit on a more equal footing with sons, reflecting current demographic shifts and welfare arrangements (Prendergast 2005). But, like Japan, there is still a long way to go before such legal definition could be fully translated into the practice of Korean families. Another point is the rights of primary carers to inheritance. Since the financial responsibility of adult children towards their parents is clearly written in Korean law, the same reform moved to grant primary carers a right to claim an extra share in return for their contribution. In this context, care and inheritance are closely intertwined and form part of the legal contract between generations. In Japan, however, Japanese law currently undermines the contribution of family carers towards the accumulation and maintenance of family wealth, except for the inheritance of *market property* (for example, family business), where those involved are entitled to receive a proportion of the bequeathed estate which will ensure its efficient working. The provision of long-term care is not legally recognized as 'earning' extra entitlement (for more discussion, see Izuhara 2002).

French law resembles the East Asian model in terms of favouring the direct line and acknowledging the reciprocal element of family obligations; however, inequality between sons and daughters in East Asian societies has a firm root, with some regional variations. As Twigg and Grand (1998) argue, inheritance in France is governed by concepts of family solidarity and reciprocity parallel to family obligations. By contrast, England presents a different model.

The dominant principle is testamentary freedom in England and in countries such as the US, Canada and New Zealand, whose legal traditions descend from that of English common law. In its pure form, individuals are free to leave their assets without consideration of rights or obligations towards their families, although a series of modifications (in 1938, 1966 and 1975) have been made to protect dependent family members of the deceased (Finch 1997; Twigg and Grand 1998). This system may allow more variations in the disposal of assets, since more kin categories, such as siblings, grandchildren, cousins, steprelatives and in-laws, are often beneficiaries. In reality, only a minority in England leave a will – approximately 30 per cent who die aged over 18 leave a will admitted to probate (Finch et al. 1996). With the increased number of individuals who own substantial assets including housing, the rate is expected to increase, however. In fact, when a person dies without a will, a similar principle to the Japanese system is applied to English cases. In such circumstances, a surviving spouse has the strongest claim, but they are not entitled to the whole estate. If the deceased had children but siblings or surviving parents exist, they are also entitled to a share (Finch 1989). In reality, therefore, despite the defined freedom, most inheritance is passed on in the direct line onto spouses and children (Finch and Wallis 1994).

The inheritance rights of spouses and children, in other words, by marriage contract or by blood ties, continue to be a major debate in some countries, including France. Under the Napoleonic civil code, for example, spouses – bound by marriage contract not by blood ties – had practically no rights to inherit since property was seen to be passing down the blood line (Finch 2004: 170). And according to Finch (2004), even in a country so progressive as the Netherlands, whether children's rights should essentially continue to take precedence over those of the surviving parent or step-parent, is still a well-debated point.

Moreover, it is worthwhile examining how assets are divided among adult children in practice. This aspect illuminates East–West differences. In England, the principle of equal shares among children is strongly held, with no preference according to gender, birth order, closeness to parents or involvement in care giving (Munro 1988; Finch et al. 1996; Twigg and Grand 1998). In Japan (and also Korea), however, 'which adult children inherit what?' by gender and birth order remains a crucial question, reflecting the traditional patrilineal succession system. Although the new civil code defines children's equal rights on inheritance regardless of their gender and birth order, they do not necessarily inherit family wealth equally. Even in postwar Japan, the larger the family size, the lower the chance that every child inherits equally. Sons, especially the eldest, are still in the strongest position to inherit the family property intact, and the typical beneficiaries are still most likely to include co-resident and married children, and sons more than daughters.

Finally, there are gender differences in the types of assets that children receive in Japan. Under the pre-war civil code women were excluded from inheritance as well as asset formation, and such gender discrimination has been perpetuated. Consequently, sons are still more likely to inherit property, and daughters to receive cash gifts, sometimes at and sometimes shortly after their marriage, thereby substituting for inheritance. Family symbols such as graves and Buddhist altars also tend to be passed on to the eldest sons. For example, of female respondents aged 61–65 years, 24 per cent inherited property and 29 per cent graves or Buddhist altars from their husband's parents, while 19 per cent inherited savings, stock and shares from their own parents (Tokyo Women's Foundation 1997). The disadvantaged position of Japanese women in asset formation reflects their economic position in the family and the labour market. An imbalance in the division of inheritance among adult children may be inevitable in some societies where residential property tends not to be defined as a tradable commodity. The debate on the meaning of home – is it a commodity, an investment vehicle or 'family assets'? – will be explored further in Chapter 5.

Institutional context: Inheritance tax

Institutional arrangements often influence people's decisions on how, when and to whom individuals' wealth is passed on.

Although more people may be becoming subject to inheritance tax as inflation pushes the level of their housing capital up, those who pay large amounts of inheritance tax are still the minority in many societies. For example, under the current tax system, in Japan only 10 per cent of those aged 45 and over incurred inheritance tax on their estates. Despite the nine-stage progressive taxation, which taxes higher asset owners much more, the bottom line of eligibility is set very high (¥90 million = £450,000 [£1 = ¥200] for the family of a spouse and two children), leaving only 10 per cent of asset transfers currently taxable (Ministry of Finance 2001). The UK has a different system, and under the Brown administration in 2007 the threshold of the taxable value of an individual's estate doubled from £300,000 to £600,000 for married couples: inheritance tax (currently 40 per cent) is payable on the excess above the threshold. However, only 6 per cent of estates paid inheritance tax in 2003/4 (when the threshold was £285,000). Unfortunately people's knowledge of inheritance law and tax is generally poor, which may be creating false alarm among the population (Rowlingson and McKay 2005).

A tax regime can indeed manipulate the ways in which individuals' assets are passed on. A good example is France, in which a regime in inheritance tax strongly favours the direct line of wealth transmission. Individual inheritances are taxed according to the nature of the relationship between the testator and the beneficiary: The average tax is less than 10 per cent in the direct line (for example, children), but 49 per cent in the indirect line such as nephews and nieces and 56 per cent for those who are unrelated (Twigg and Grand 1998).

In order to avoid punitive inheritance tax, some people make *inter vivo* transfers by taking advantage of tax-free allowances, or they cash in the property when the value is high, so as to leave as many assets as possible to their children when they 'need' them. *Inter vivo* transfer is more likely to occur under the current Japanese system of generous tax-free cash allowances, in particular to help children for house purchase or improvement.

Housing assets and social policy

As part of the global trend of welfare retrenchment, politicians and policy makers have increasingly seen individuals' assets as a source of income in old age. Whether people accumulate assets for future generations or to improve their own living standard in old age poses an interesting question. The asset-based welfare approach of using personal assets to fund the cost of care is, however, relatively new to social policy, given the fact that the experience of owning assets is rather new to the majority of the population in some societies (see, for example, Finch 2004). Thus, as Finch (2004) suggests, the attitudes may not be well formed or fixed in society and there may be clear cross-national differences.

How material assets are seen across generations also varies within and across societies. The dominant cultural norm in contemporary Britain is, for example, that children *do not expect* inheritance. Adult children may be given something on the death of their parents but they do not usually 'count on it'. Neither is there evidence that children are seen as having any 'rights' over their parents' property (ibid.). Rowlingson and McKay's survey (2005) on people's attitudes towards inheritance also confirms this point that inheritance has not become entrenched as an expectation or duty, and most older people are willing to use their assets for themselves to meet needs in old age. In other words, for those new generations of home owners, 'material assets are treated as the property of one generation rather than as the foundation of wealth in future generations' (Finch 2004: 172).

Such individualistic understanding of household assets may be contrasted with the traditional notion of wealth embedded in the family system in East Asia. However, at the same time such attitudes contradict the popular discourse among older parents today in the English context. Some feel betrayed by the government because, having paid taxes all their working life, they are then charged for the costs of care largely because they want to protect the assets of younger generations. The debate about the social contract appears to be somewhat divorced from the rights and responsibility of the family. Japanese views are somewhat different and again do not really support the traditional ideal of transgenerational asset accumulation of family wealth. Hirose *et al.*'s survey (1998) confirmed the

Housing assets and intergenerational transfer

willingness of the first generation of home owners to leave housing assets to their children. The importance of the family was certainly behind their motives, but not necessarily part of the reciprocity derived from the continuous chain of the 'generational contract' as many of those people achieved home ownership entirely by their own means (income from labour market participation) rather than by inheritance.

Strategies of using housing assets in old age

So far this chapter has focused on one of the main options available in disposal of housing equity: passing it on as inheritance after people's death. This final section briefly explores other options available to older home owners such as whether to withdraw or extract housing equity and use it for their own purpose. Policy makers and some academic commentators increasingly view turning existing real estate assets of individuals into an income stream as a necessary strategy in the revitalization of the economy and also in asset-based social policy (Izuhara 2007). There are many options available worldwide on turning housing stock into an income stream to help older people provide more financial security in their later life. The development of various means can indeed be inconsistent, depending on the development of certain national conditions such as financial markets and housing markets.

For 'asset-rich, income-poor' home owners, reverse mortgage schemes are in theory very attractive options for those who have a limited income and would like to release some of their equity tied to an owner-occupied house while remaining (semi-)independent in their own home. The developments in financial markets are widening the options open to households for using housing assets. These developments constitute both a threat and an opportunity in the intergenerational context.

Trading down involves both inter and intra-tenure downward movements in terms of size and value of property. Older people can move to purchase a smaller, more manageable and less expensive property to release some equity from current housing. This is a common approach for empty nesters in the West. However, the housing market, size and location of properties highly dictate how much wealth older people can release and thus the destinations of their moves. For older people moving to purpose-built housing is one of the options available in some societies – it is a desirable option for some who want to remain semi-independent longer. An inter-tenure movement from owner-occupation to tenancy can be attractive for others who view owning a house in old age as a liability in terms of maintenance and property tax. However, if people rely on the 'risk-averse' rented housing market in old age, problems may arise or be accentuated when they reach a certain age.

Furthermore, renting out additional properties is one way to achieve extra income given the increasing possibility of housing inheritance in later life.

The recent deregulation of the Leasehold Act in Japan, for example, opened doors to this option for non-professional property owners and investors. However, this option may be available for only a fortunate few who have invested earlier or inherited residential properties. Increased inheritance prospects in a low-fertility society may expand the beneficiaries of this option, but again the location, age and type of housing stock will dictate the rentability and thus the level of rents achieved. As the number of vacant properties is already becoming an issue in contemporary Japan, winners and losers may be divided by how their additional properties meet the demands of the changing rental market (Izuhara 2007).

Conclusion

The growth of home ownership in many capitalist societies meant increased prospects of inheritance in families, especially when the first generation of home owners entered old age. The current socioeconomic and demographic changes may, however, complicate people's expectations and experiences. This chapter looked at the role of home ownership in the accumulation of family wealth over generations and how such intergenerational transfers impacted on family relations. Inheritance is important in shaping family relations since it tends to give an advantage to family members. However, while it contributes to family solidarity, at the same time it may well become a source of conflict.

The processes of wealth accumulation and inheritance outcomes are not homogeneous but indeed have marked cross-national differences considering the differences in culture, market conditions and institutional context. Across societies, factors such as economic boom and bust, institutional mechanisms, available housing stock and ownership status could contribute to differentiated processes and outcomes. At a societal level among individuals and families, factors such as gender, family structure, timing of purchase, type and location of properties dictate the inheritance prospects and outcomes. Asset-based public policy, including the development of home ownership, can be seen as processes of social stratification.

The rhetoric associated with an asset-based welfare state is gaining popularity under the current global ageing and economic pressure. In this context, in many advanced economies politicians and policy makers have increasingly viewed individuals' housing assets as a source of income in old age. This is closely linked to the debate regarding the cost of long-term care in an ageing society. Despite the global trend of welfare retrenchment, the Japanese government has formulated a contrasting strategy under the new social insurance scheme on long-term care, where an individual's income and assets are irrelevant in the assessment process to receive public services. The next chapter explores this policy issue, in particular the interaction between such micro-level family relations and the wider social structure and policy on the delivery of long-term care. It examines how the different generational contracts have been shaped by the different policy logic of a nation.

4 Long-term care and shifting state–family boundaries

Introduction

Policies on welfare provision tend to draw a boundary between the state and the family, but such a boundary is highly contested and shifts over time according to changes in a nation's social and economic circumstances, demographic patterns and political agenda (Fox Harding 1996). This chapter aims to explore how in relation to the responsibilities of long-term care for older people the state–family boundary has shifted in contemporary societies in the East and the West. The generational contract, or intergenerational relationship within the family, which is the main analytical framework adopted in this book, forms the backbone of a wider social contract defined in each national context. It is misleading, however, to divorce links between the different levels of 'generational contracts' since they are often mutually influential (Walker 1996). This chapter thus examines how in the context of broader trends in the political economy state–family boundaries are shaped and reshaped by policy logics and various welfare reforms.

With the power of law and public policies, the state is often a facilitator in drawing such boundaries by encouraging (or discouraging) certain family behaviour. On the other hand, the state can also be a follower of trends and implement some measures to support and confirm already-shifted boundaries. Although governments may be ambivalent about the boundaries of responsibility in social care provision, partly owing to tasks being carried out in the domestic sphere, the family is usually 'expected' to have the main responsibility. In many respects, even in societies with highly sophisticated public services, formal social services have been expected to come into play only when family care fails (Hill 1996). Public policies indeed frame the context of family obligations in a number of ways: defining the role of the family and forcing them to perform certain tasks not only through law and clear policy measures but also by legislating intergenerational obligations (for example, the Maintenance of Parents Act 1994 in Singapore); providing benefits, assistance and incentives such as tax concessions; and also through the absence of alternative, 'appropriate' and adequate support and services. Howe (2001) defines the differences in balances of approaches in the Pacific Rim nations as: (1) redefining the relative limits of family care (by realizing the hidden potential for family care through legislation or the inclusion of cash benefits in programmes and supporting families in normative care-giving roles); or (2) dealing with an absolute limit (substituting for the absence of families).

As Finch (1989) states, 'any government wishing to restrict public expenditure is likely to explore how family ties can be strengthened either explicitly or implicitly by defining such obligations'. Others argue that governments' emphasis on traditional family values held back the development of some areas of welfare programmes. This is particularly the case in East Asian societies where Confucian values emphasize the role of the family and the moral duty of adult children (Ikels 2004; Tao 2004). In Japan, for example, compared with major items of public expenditure – pension schemes (52 per cent of total expenditure in 1997) and national health insurance (37 per cent) – social service provision (11 per cent) has been considerably underdeveloped until recently (National Institute of Population and Social Security Research 2000 www.jpss.go.jp). There are various explanations for such modest involvement of the state in this policy arena: first, there is a common belief (also backed up by political ideology) that the family is a private institution and the care needs of older people (as well as those of dependent children and disabled members) are considered to be a private matter (see, for example, Shin and Shaw 2003). Second, the notion of developmental states in East Asia is particularly apparent in this policy field as social policy subordinates a nation's economic growth (Holliday 2000). Third, in some societies the state plays a predominantly regulatory or coordinating role rather than providing direct services. Moreover, there is the 'substitution' hypothesis that the generous provision of formal services to support older people 'crowds out' family support, and the provision of care predominantly falls on to the responsibility of the state (Motel-Klingebiel *et al.* 2005). All these reasons tend to inhibit effective policy development and result in underdeveloped formal services. The comparative project in Europe, however, did not confirm the substitution hypothesis (Motel-Klingebiel *et al.* 2005). Their analysis shows that the total quantity of support received by older people is greater in welfare states with a strong infrastructure of formal services, and no evidence of a substantial 'crowding out' of family support was found. Instead, it concluded that in societies 'with well-developed service infrastructures, family support and public services act accumulatively, but that in "familistic" welfare regimes, similar combinations do not occur' (Künemund and Rein 1999; Motel-Klingebiel *et al.* 2005).

In more recent years, however, the emergence of a 'new social risk' – an increase in the long-term care need to provide for older people – in many advanced economies has turned a 'private issue' into a public concern (Morel 2006). For the past decade, the social crisis over demographic and gender shifts,

Long-term care and state–family boundaries

or, as Peng (2002a) calls it, the 'public crisis of the family', is evident in various national contexts. The fact that the reforms on long-term care have been brought into the spotlight in many societies is not only because of rapid demographic change (population ageing) *per se* but also largely due to other socioeconomic and political factors (Chi *et al.* 2001). Those factors shed light on such contemporary debates by questioning the feasibility of developing as well as sustaining the existing state–family boundaries.

Although the focus of this chapter is given predominantly to the relationship and interaction between the state and families, the fact that there are various other actors/sectors involved in this arena should not be ignored. The mixed economy of welfare is increasingly promoted under policy reforms, and both non-profit and for-profit organizations now play an important role in the delivery of social care in many societies. Indeed, one of the aims of the new social insurance scheme in Japan was *socialization of care*, expanding both services and providers beyond the conventional state–informal sector mix with the introduction of a market mechanism (Hiraoka 2002).

This chapter aims to set the wider theoretical framework, considering trends and developments across OECD countries. And the comparison goes beyond the welfare models of Western societies by including East Asian societies, so that the cases of Japan and Britain stand out. It will also aid analysis of data on micro-level interpersonal relationships drawn from the empirical research discussed in following chapters. This chapter also explores whether or how the East–West comparison is particularly helpful to highlight and answer the boundary shifts on long-term care. Finally, owing to the informal nature of family care and tasks being performed in the domestic sphere, formalizing and monetizing the quantity and quality of informal support are very difficult and thus difficult to compare across nations (Hill 1996). The comparative data used in this chapter therefore need to be interpreted with caution owing to the variation of national definitions and statistical categories.

Key drivers of change

There are considerable demographic and gender shifts occurring across OECD nations as well as among East Asian countries that are influencing existing state–family boundaries. The ways in which they are shifting have various national characteristics within shared global trends.

As discussed in Chapter 2, population ageing caused by falling birth rates and prolonged life expectancy has been a common policy concern of the global North. Older people are a diverse group with different health, financial and family backgrounds, and age alone has a different significance in terms of policy formation and forecast. The growing rates of the young-olds are, for example, not necessarily an alarming factor given the increased level of their economic participation and productivity, and also due to their frequent contribution as care givers both to their partner and elderly parents. The growth of the oldest-olds aged 80 and above is, on the other hand, an important issue since this fastest growing cohort will increase demand for long-term care and threaten the sustainability of future costs of care, thereby increasing the perceived burden on family care. Table 4.1 provides a comparison of the proportional growth of this age group across selected OECD countries. In parallel to the rates of societal ageing, there is a clear divide between mature Western societies where growth in the rates of societal ageing has slowed down noticeably (for example, Denmark, Sweden, UK), and the fastest ageing societies where, with the intense growth of the cohort, the issues will become more pressing within a shorter space of time (for example, Japan, Korea and Poland).

According to the OECD report (2005), informal care currently provides over half of all long-term care in the selected countries studied, and the proportion may be greater in the East owing to a lack of alternative services. Given the assumption that families do provide care, the capacity of families to sustain the existing and expanding care load in the future is another side to the discussion, and there are a few comparative points to scrutinize.

Table 4.1 Share of very-old persons (80+) among older people in selected OECD countries, 1960–2040

				Change in % points	
	1960	*2000*	*2040*	*1960–2000*	*2000–40*
Australia	14.3	23.6	31.8	9.3	8.2
Austria	14.4	22.8	28.1	8.4	5.2
Canada	15.8	23.6	32.9	7.8	9.3
Denmark	15.3	26.7	28.9	11.4	2.2
Finland	12.7	22.5	35.1	9.8	12.6
Germany	–	22.3	29.9	–	7.6
Ireland	17.5	23.0	26.7	5.5	3.7
Italy	14.6	22.2	30.6	7.6	8.4
Japan	12.6	22.0	41.1	9.5	19.1
Korea	8.1	14.2	26.1	6.1	11.9
Poland	12.2	16.2	31.9	4.0	15.7
Sweden	15.9	29.0	31.5	13.1	2.5
UK	16.4	25.4	29.1	9.0	3.7
USA	15.2	26.4	33.3	11.2	6.9
OECD average	14.4	21.7	30.1	7.3	8.4

Source: OECD (2005: 102); 1960 and 2000: OECD Health Data 2004; 2040 projections: Eurostat (EU countries); national sources (Canada and US): United Nations (2002).
Note: Germany 1960 (before reunification) is not comparable with 2000 data.

Howe (2001) highlights the dichotomy of the development of formal service delivery in relation to changing families in the East and the West by questioning the compatibility between increasing professionalism and the promotion of family-centred home care (Chi et al. 2001). In East Asian societies, 'family breakdown' is more apparent in the household structure as family nuclearization among older people has accelerated (63 per cent in Hong Kong in 1996 and approximately 50 per cent in Japan in 1998). Marriage is still a popular institution but there has been an increase in divorce and never-married people. In those societies, changes in traditional values regarding family support for older people provided a rationale for developing policies *to support family carers*. On the other hand, the level of family care has been consistent in the developed Pacific Rim countries of Canada, Australia and the US, and formal services are often sought by those without families (Howe 2001). Another difference in demography is that much larger numbers of older people, especially women, have never married or remained childless in those countries. Many commentators (see, for example, OECD 2005) predict that older people living alone without a resident partner or younger family members will increase demands for formal care services in the future. There are also global policy debates around de-institutionalization which tends to add to the cost of care since (semi-)independent living at home can be more expensive than care provided in institutions (Parker and Clarke 2002). The breakdown of marriage and partnership, and subsequent re-partnering, which are increasingly common in contemporary societies, jeopardize 'expected' family support and make obligations less clear within the family and across generations. Overall, the different shifts in households between the East and West produce interestingly similar outcomes of reduced caring capacity. The decline in co-residency has been reducing the sources of informal carers in the East since children tend to be the main carers (60 per cent in Japan and 55 per cent in Korea in 2001; see OECD 2005: 109), as does the further increase in divorce and family breakdown in the Western societies where partners often have the main caring responsibility.

Furthermore, the state–family boundary is contested but it is highly gendered. It is consistent across societies where women shoulder a substantial part of the burden of informal care for older people with long-term care needs. According to the OECD (2005: 109), over 70 per cent of carers were women at the turn of the new millennium (Australia: 71 per cent, Canada: 73 per cent, Germany: 80 per cent and Japan: 76 per cent). They are more likely to be main carers rather than additional carers, and tend to engage in the heavy end of care giving. Women's increased participation in formal labour markets for a variety of reasons is nowadays well documented, signifying the declining capacity of informal care. OECD labour force statistics (2005: 111) indicate that such an increase has been more significant in some 'conservative' societies for the past two decades (for example, Spain, the Netherlands and Germany) than more 'liberal' societies (for example, the Nordic countries where the rates have been high since the 1980s) and 'traditional' Eastern societies (for example, participation is still lower in Korea and Japan). Within a peak age for care giving (45–65), a substantial increase was reported among the 'generation in the middle' – those aged between 45 and 54. This may be something to do with the current global economic climate requiring a contribution of women (additional earners) to the household economy, due partly to increased costs for housing and children's education. The international trend towards a 'dual-worker model' (as opposed to the traditional male-breadwinner family model) is a reflection of the contemporary labour market structure, or the 'one-and-a-half worker model', if we consider the fact that women are often engaged in part-time work. Part-time work allows women to perform other tasks in the household such as care of dependent adults and children, but also at the same time creates a dual burden of work and family life. This shift of women moving from unpaid to paid work, or managing both paid and unpaid work, needs to be taken into account when examining state–family boundaries. Recent studies have, however, suggested the dichotomy between paid and unpaid work is diminishing in care work across developed welfare states where informal care is increasingly 'commodified' (Ungerson 1997).

Another interesting point concerning women and care giving is migration in the global economy. There are several aspects here. First, the caring capacity lost in some areas, such as rural communities and some socioeconomic groups, owing to women's high mobility and visibility in the public sphere in one country may rely on internal or international migration to fill the gap. In Japan, for example, rural areas where family tradition remains strong have been suffering from de-population as well as a rapidly ageing population, yet the level of public services tends to be lower. Changing families in this context mean the difficulty of finding brides (*bride famine*, as the media portrays) for successor sons in farming families, who traditionally bear the caring responsibility. Such a struggle in rural farming communities led to the emergence of a new industry – first, the municipal intermarriage introduction service during the bubble economy period of the 1980s, and subsequently commercial matchmaking services (Umeda 2002). These services recruit brides from less developed neighbouring countries such as China, Korea and the Philippines to marry local young men. The more 'commodified' version of care supply is increasing the importance of migrant care workers to fill the gap vacated by the family. For example, the common reliance of illegal migrants to fill in the care labour market in urban Italy (Ungerson 2004), and more formally increasing foreign workers in the health and care labour sector in the EU (Harper 2005). Moreover, in a globalizing world, women – potential carers of the family – move on their own and of their own will nationally and internationally. Ackers' (1998) research on 'women, citizenship and migration' within the EU highlighted well those migrant women's

Long-term care and state–family boundaries

attitudes towards their responsibility for caring for their ageing parents left in their original country and their coping strategies over time and space.

Care and logic of welfare states

Social care has always occupied a distinctive position in the development of many welfare states, with both an explicit and implicit assumption on the role of the family as welfare producers. Informal care provided at home by family members is still the most important source of care in all countries (OECD 2005). Not only *private households* but *extended families* in and away from co-residency share the burden of care in most countries. The extent to which they provide support varies from just 'keeping an eye on' family members to helping with regular shopping and providing the heavy end of nursing care. The types of support provided include personal, practical and emotional support, providing accommodation as well as economic support (see Finch 1989). Financial contributions by families, and subsequent strains on them, for example, come in various forms such as making substantial co-payments, out-of-pocket spending for care provided under public programmes both at home and in institutions, and also the loss of opportunity costs in the formal labour market. The provision of informal care often being unpaid makes the state substantial savings, as even in countries with relatively comprehensive coverage, spending on long-term care is currently only around 10 per cent–20 per cent of total spending on health and long-term care together (OECD 2005).

A nation's long-term care programmes often go hand in hand with the policy logic of welfare states. Using Esping-Andersen (1990)'s welfare regime typologies (although criticized due to a gender and family-blind analysis and analytical framework) (Lewis 1992; Sainsbury 1994), older people have traditionally been looked after by the family, especially in 'Conservative' welfare states where the principle of subsidiarity has long prevailed (Morel 2006). The social rights of individuals are well established in the Nordic 'social democratic' welfare regimes so that dependent older people use public services more, while older people in 'liberal' regimes rely more on the market. When looking at East Asian 'Confucian' welfare states, it is often said that the strong moral values of responsibility, filial piety and respecting seniority make families provide more welfare than in the West. Many commentators therefore emphasize that economic success in East Asia is often at the expense of welfare in these 'developmental states', as the focus is on industrial development and national economic security, and the subordination of social to economic policy (for example, Holliday 2000; Peng 2002b). By examining the connection between anti-welfarism and market ideologies, however, some scholars challenge the core assumptions that underpin the claim that social welfare is unAsian because social welfare in Asia is underdeveloped. Thus the underdevelopment of social welfare contributes to

Asian societies' economic success. Furthermore, Asian values do not necessarily promote the development of social welfare (Chau and Yu 2005). Indeed, there needs to be more scrutiny of the myths or misconceptions that families in the East produce more welfare than those in the West, and that social care programmes are underdeveloped in East Asia.

Based on Chinese Confucianism, looking after older parents is often understood as part of the 'generational contract' – a moral obligation on adult children to reciprocate the care of their parents in the process of their upbringing. For Tao (2004), however, under the Confucian system of ethics, the moral basis of reciprocity is not a 'contract' or utility but rather represents interconnectedness and interdependence. And such intergenerational obligations are explicitly written in the laws in some countries, but in others they are based on consent and voluntarism. For example, many societies in the East, such as Taiwan, Singapore and Mainland China (and also in the West, including Spain and Israel) have laws that require adult children to provide maintenance for their ageing parents. In societies such as Hong Kong and Japan, in contrast, the support is largely a normative expectation but of a voluntary nature. For example, in Japan, although there is no legal obligation for children to provide for their parents financially, in practice public assistance to older people is restricted or rationed if the extended family (adult children) have a 'support ability'. In contrast, for many Western philosophers, filial obligation is regarded not as a moral obligation but based on consent and voluntariness (English 1979; Daniels 1988, cited in Tao 2004). In most cases, the nature of the caring relationship rests on a balance between reciprocity, affection and duty (Marshall *et al.* 1987; Finch and Mason 1993).

Recently in the global North there have been some shared trends in the development of social care services and providers as well as changing balances in the mixed economy of care. Many developed economies have experienced a rise and fall of the welfare state in the postwar period. Brief highlights of the significant boundary shifts occurring in different periods in different national contexts are as follows. The 1950s and 1960s were a period of expansion with greater involvement of the state in social care provision in many developed nations. In Sweden, for example, social services became a basic right for all citizens, and were paid for and organized by the state in the 1950s. In the 1960s, changing gender relations played a part in the expansion of social services, peaking with the growth of the formal services (Theobald 2003), but by the 1980s older people viewed public services as the favourite option over family care in Sweden.

In Japan, along with many other economic policy reforms, the widening scope of social policy and welfare provision was typically driven by external forces when the OECD requested a comprehensive report on social policy in the 1970s (Tamai 1997). In 1973, seeking more welfare and a better quality of life, the government moved to introduce *Fukushi Gannen* [Welfare Year One], and referred to the ambitious transition of state priority from economic growth to welfare. As a result the rate of the social security budget in the fiscal year 1973 was raised by 29 per cent. However, the momentum of the reform was immediately hit by exorbitant land price inflation, the upheaval of the international monetary system and, more significantly, the first oil crisis. In this period, the government's commitment to further expand state welfare was jeopardized and economic and political priorities were altered. The slogan of 'Welfare Year One' was gradually replaced by 'reconsider welfare', 'welfare state disease' or 'Japanese-style welfare state', stressing a traditional Japanese spirit of self-reliance. Since then, the family role for social care was redefined in the late 1980s under a series of Gold Plans, emphasizing a 'home care' option with an expansion of various community care facilities and services in response to an ageing population and changing families. The more significant turning point came in 2000, however, with the introduction of the new social insurance scheme on long-term care. The scheme created a 'quasi-market' in which alternative sources to families (both non-profit and for-profit) compete to provide services to those in need.

In the UK, following the global economic recession of the 1970s, the 1980s saw cuts in social welfare spending, shifting much of the state burden back to the family. Such a New Right approach under the Thatcher government became a global phenomenon, emphasizing the restriction of the role of the state in economic life, and focusing on market forces as the basis of both individual liberties and economic growth. As Ackers (1998) states, increasing reliance on the informal sphere is the corollary to welfare retrenchment. Significant reforms to privatize and marketize welfare provision, together with changes to the structure and operation of state welfare services, took place during the 1980s. In social security, for example, targeted (means-tested) provision such as Income Support replaced the universal (but contributory) provision of National Insurance benefits. A shift from the direct provision of public services and facilities to private market provision supported by means-tested public subsidies was also evident in residential care for older people. The Thatcherite principle of a free economy and a strong state also re-emphasized the role of the family in welfare provision.

Increasing emphasis on community care under the National Health Service and Community Care Act 1990, backed by rhetoric about 'families wanting to care', meant a cutback in the delivery and availability of support services, and a reduction in the availability of residential care, hospital care and mental health institutions (but ensuring older and disabled people were able to exercise the same choices about where and with whom they lived as anyone else). There are anxieties about the foreseeable cost of an ageing society with an increasing number of 'dependent' older people who may be a drain on resources. At the same time, there is an ideological commitment to individual responsibility for personal welfare, and this has resulted in more pressure on families and the community as 'proper' providers. Dealing with a combination of different issues

Long-term care and state–family boundaries

in a single policy arena, as Parker and Clarke (2002) argue, only creates problems and has made it almost impossible to ensure that older and disabled people are able to live in their own home unless they are supported by unpaid or unsupported informal carers.

From the 1990s onwards, there has been an international trend towards a more mixed economy of providers. For example, Sweden introduced a market principle with a more mixed economy of both providers and services, while countries such as Germany and Japan opted for social insurance schemes. In many welfare states, however, the family is still viewed as the central player in the provision of social care. For example, the German scheme provides only basic benefits, leaving care recipients and families with the main responsibility (Theobald 2003). This is in contrast with Sweden, where social care is mainly seen as a societal obligation and access to social care is defined universally as a citizenship right. The low level of benefits also reinforces inequalities in societies such as Germany.

As the 1990s have been considered as a period of retrenchment, the creation of such insurance schemes in those Bismarckian welfare states is puzzling to some scholars (Pierson 2001, in Morel 2006: 226). The rationale behind the reforms is due not only to actual and perceived socioeconomic and demographic shifts but also to the political agenda of the nations that influenced policy direction (Peng 2002a). Unlike Japan, some countries claim that changes in family patterns and female preferences have not necessarily been the key drivers of reform. Despite the introduction of schemes in France and Germany, for example, figures for informal care have remained the same (Morel 2006). The main difference now is that those informal carers are paid, albeit not at full market rates. In this scenario, the boundary has not shifted in terms of actual labour (support provision) but in terms of its cost burden.

Options for finance, provision and location

Over the past decade some countries have undertaken major reforms on long-term care provision. The key policy directions, which are shared globally, are: (1) shifting the balance of care towards home care, signified as de-institutionalization, community care, and ageing in place; (2) shifting the mixed economy of care by expanding alternative providers from the conventional state–family nexus; and (3) expanding care options to include cash benefits to support, compensate and reward informal carers. What are the impacts of such policy directions on families?

Long-term care reforms (options for finance and provision)

How the costs of long-term care can be shared between the state and individuals (and families) is an area of critical policy debate in many developed economies today.

There are a variety of institutional mechanisms to finance and provide long-term care (Royal Commission on Long-term Care 1999). Over the past decade some countries have developed their pre-existing tax-based systems (for example, the Nordic countries and the UK), while others, such as Germany and Japan, undertook major reforms, introducing a social insurance scheme. Others, including the Netherlands, have a hybrid of the two approaches, financed by social insurance but with services provided by the social welfare system. The two main approaches differ not only in the mechanisms of funding, but also in coverage, eligibility and entitlement (Ikegami and Campbell 2002). There are pros and cons in these approaches; for example, a tax-based model is more flexible in providing benefits according to an individual's need, since income (and assets) levels and the family's ability to provide care are taken into consideration. But a means-tested approach narrows the section of society to be recipients. On the other hand, a social insurance model can be more rigid since an individual's rights are more explicitly defined (ibid.). However, it creates universal coverage within the category of those covered by the scheme, regardless of their income or family backgrounds, and is also likely to provide more opportunities for choices instead of an 'allocation' of services.

In terms of provision, despite the common welfare retrenchment among OECD countries, there is a trend towards more universal public provision instead of targeted services provision. The complexity of long-term care lies in the fact that the responsibility of care is ambiguous (and thus public–private boundaries are ambiguous) since there is no clear location or providers of care (ibid.). Care can be provided either in institutions or at home, and types of carers and care arrangements vary from the 'classic' informal carers (unpaid, unsupported, unregulated) to care workers recruited through the labour market and subject to contractual relations (Ungerson 2004). As Ungerson (2004) argues, the policy and labour market contexts influence the precise position of each group of carers and their positions can shift with changes in policy and the labour market. Moreover, the conditions of individuals needing care may change over time, which requires constant adjustments in the public–private boundaries of care provision.

A great variety of packages exists in terms of financing and provision even within those countries currently taking similar approaches. For example, the Japanese scheme draws funds from a multitude of sources (roughly half by designated insurance premiums and another half from both national and local tax revenues) and provides in-kind services only. The Austrian scheme is funded fully by taxation and only provides cash benefits. The German and Luxembourg models raise funds from insurance premiums and provide both cash and in-kind services (OECD 2005). Differentiating types of support and funding only specific elements are another way of controlling public expenditure. In England and Wales, despite the recommendation made by the Royal Commission on

Long-term Care in 1999, only nursing care (not personal care) provided in institutions is currently free of charge. The redefinition of health needs as social needs is highly contentious in the scheme. In many societies hotel costs are often charged to residents in institutional care.

Tax-based models have often developed from pre-existing systems of social services targeted at low-income older people, by gradually expanding the services and coverage. However, there is a clear divide within those countries. Some, such as the Nordic countries, have achieved more universal coverage, while others, including the UK and Australia, still use a means-tested approach. In other less developed OECD nations such as Hungary, Korea and Mexico, public funding for long-term care is still relatively low (OECD 2005). These countries inevitably provide a limited amount of care in institutions and lack the provision of home care to support or replace family care.

Among those that have opted for social insurance, the Japanese scheme proposed that caring responsibilities should be shared in society with a variety of new providers, instead of leaving responsibility entirely with families. The scheme has been successful in expanding the number of services and providers, and has begun to include a much wider section of society as recipients, in particular those who used to be excluded from public services owing to their financial and family circumstance. Before the long-term care insurance scheme, social services were means tested and the income and 'availability of family resources' were some of the criteria preventing access to public welfare. At the time of introduction, socialization of welfare was partly successful in shifting state–family boundaries. The cost burden has shifted for some people who used to pay in full out of their own pocket, but others, particularly those on a low income, may opt out of the entitled formal services due to the double burden of paying for insurance premiums and user fees (see Izuhara 2003b).

Moves to develop an independent insurance scheme (away from health insurance) are often opposed owing to the projected cost implication. Interestingly enough, however, whichever model the nation adopts, so far it has led to similar expenditure outcomes. As Figure 4.1 indicates, total expenditure on long-term care in the selected OECD countries is from below 0.2 per cent to close to 3 per cent of GDP (gross domestic product). Despite their different approaches, Australia, Canada, the US, Germany and the UK have similar expenditure rates. Most countries are clustered in the narrow range between 0.5 per cent and 1.6 per cent, with few exceptions. Norway and Sweden (tax-based, universal coverage) have an expenditure ratio well above the average, while the Spanish rates are below 0.2 per cent (data is not available for Korea).

In many societies public expenditure on long-term care as a proportion of GDP is projected to double by 2050 (OECD 2005; Comas-Herrera *et al.* 2006). Sustainability is indeed a key issue for those countries that introduced radical reforms to this highly ambiguous policy field, with contested boundaries between

Long-term care and state–family boundaries

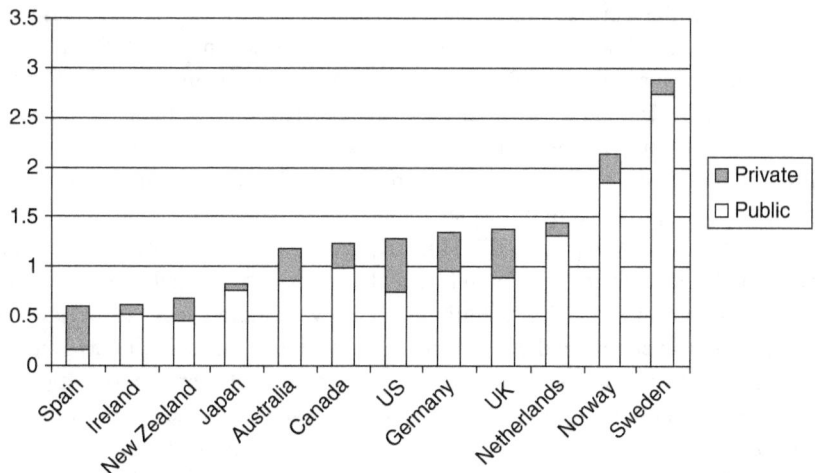

4.1 Public and private expenditure on long-term care as a percentage of GDP (2000).
Source: OECD (2005: 26).

the public and the private; between health and social care, and also between home and institutional care. In the Japanese scheme, at the first five-year review in 2005, financial sustainability was already an issue. Owing to the traditional stigma attached to 'public welfare' and also the private nature of domestic arrangements, there was anticipation at the beginning as to whether the current older generation would take up the services. The ideological change soon took shape, and the take-up rates for services increased significantly over the first few years, especially the waiting list for nursing homes. People now view receiving public services as their entitlement and as an integral part of long-term care provision in society. As a result, the five-year review proposed an increased cost that would have to be shared in three ways: an increased premium, raised co-payment and wage-cut for care providers. It also abolished domiciliary services for those who were granted the lowest need categories (support need 1 and 2). Indeed, the review redefined the role of family carers by re-introducing the 'availability of family resources' as an assessment criterion to cut back services (Hiraoka 2006). This is a step backwards since the uniform entitlements (in return for mandatory contributions) were one of the key features of the new scheme.

Location of care: Care home or institutions

The location of care giving often determines state–family boundaries. People's preference for staying put has been widely recognized in many societies and is supported by the current policy direction of promoting home care. 'Ageing in place' has indeed been a part of the global agenda since the mid-1990s and the general policy direction in OECD countries is aiming to maintain frail older people in their own home for as long as possible. The starting point from which the balance of home and institutional care is to be established is, however, very different in different national contexts (Chi et al. 2001). Even within Europe, the balance of care provided by informal carers, formal home care or in institutions varies widely. For example, in 2000 the majority of long-term care (69 per cent) was provided by informal carers in Spain (14 per cent by formal carers at home and 17 per cent in institutions) while the rates were more evenly distributed as 32 per cent, 36 per cent and 32 per cent respectively in the UK (Comas-Herrera et al. 2006). In mature welfare states, the promotion of community care may mean 'de-institutionalization' – a reduction of already available care institutions and hospital places. On the other hand, in other societies, including East Asia, where only a significant minority of older people are cared for in institutions (for example, 2 per cent in Singapore), it means by-passing the processes of developing such care facilities. In those societies, the provision of home services tends to be equally low (for example, in 1998 less than 1 per cent of older people received home care services in Hong Kong). The scarcity of residential homes and nursing care institutions in Japan, for example, was substituted by hospitals and medical clinics, resulting in growing deficits in medical insurance. One of the motivations behind the Japanese reform was therefore separating long-term care from medical care with the expansion of services and facilities, and the introduction of tighter systems of co-payment.

As Figure 4.2 indicates, the majority of public expenditure spent on care was on care provided in institutions. This is partly because the majority of those in need who live at home maintain themselves not only through receiving public services alone, but often with the support of informal carers, friends and neighbours. If frail older people remain at home as long as possible, care is generally provided by family members and it is often hard physical and emotional labour (Ungerson 1994). Such informal carers are often unpaid or underpaid, which makes home care a lower cost alternative to institutional care (OECD 2005). Although community care has often been presented as offering a lower cost alternative, as Howe (2001) argues, a more diverse array of cost pressures have been identified. These could come from a shift in functions from healthcare to long-term care (as, for example, in Canada), a mismatch of increased expenditures by both the state and individuals and increases in quality of care (as, for example, in Australia) and anticipation of cost increases blocking policy development (as, for example, in the US and Hong Kong).

Even in those countries that have more developed public services in this area, close relatives feel they are the people with whom the final responsibility rests, and as Ungerson (1994) argues, this creates a complex relationship between

Long-term care and state–family boundaries

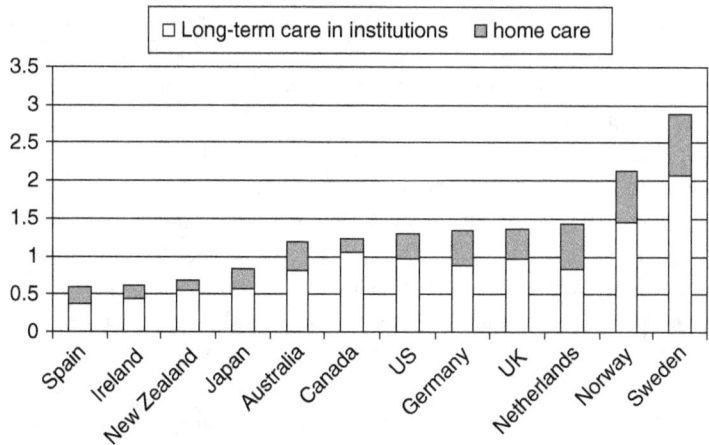

4.2 Home care versus institutional care – public expenditure on long-term care as a percentage of GDP (2000).
Source: OECD (2005).

carers and the cared-for. Recognizing the need of carers has gradually become a policy issue in recent years and many OECD countries have adopted a number of policies to support carers. The range and level of services available to informal carers vary across nations: the UK, one of the progressive nations, has now given carers a statutory right to receive an assessment of their need for services in addition to services for those being cared for, and also pension credits to enable those out of the labour market to maintain pension rights, such as the Netherlands, Australia, Canada, Sweden, Germany and the UK (Evers *et al.* 1994; Glendinning and Kemp 2006). Among those countries, the Dutch and Austrian systems represent examples of the fully 'commodified' informal care as family carers are paid in cash according to the amount of care work time, and covered by social security rights, holiday and sickness pay (Ungerson 2004).

Care options: Services or cash benefits

Linking the debates on consumer choices and the issue of carers, care options present interesting and relevant points for analysing the boundary shifts. A key question arises though: how do choices (or the availability of such choices) impact on family carers? Do they relieve the burden of care from families or do they reinforce the role of the family? Over the past decade, a variety of programmes have been developed across OECD countries allowing older people and their families to practise individual choice among their care options.

Long-term care and state–family boundaries

For older people more options can often mean a greater degree of control over their lives, and such choices and consumer direction are likely to contribute to a better quality of life when adequate services are also provided to support their carers (OECD 2005: 12).

International developments in this area are moving towards cash benefits with some exceptions, such as Japan (see, for example, Evers *et al.* 1994; Glendinning and Kemp 2006). Payments are made through various channels, including the funds being given to the individual needing care, who is then responsible for purchasing the required services in their own home. Frail older people are thus being given the means to enter the labour market and to contract caring labour ('routed wages') (Ungerson 1997). To compensate for expenses incurred due to care giving, or on the basis of an agreement to provide a certain level of services, funds may be given to carers, directly or through the fiscal system, or though contributions to a pension plan (Stryckman and Nahmiash 1994). For Stryckman and Nahmiash (1994), payment for care can be seen as: compensation (reward or recognition) to informal carers, giving clients greater control over services (instead of the conventional allocation of care by the public), as job creation, as a policy to get women out of the labour market and back into the home, as a form of privatization, or above all, as a cost-saving measure for governments who are facing budget restrictions.

In some countries with a high level of service delivery, people may fear that an increase in payments for care will offset existing formal services (and thus the physical and emotional burden will remain within the family). The experiences of various OECD countries demonstrates that there is a mixture of outcomes, depending on the state's commitment, resources and the chosen systems (Ungerson 1994). Despite the introduction of cash benefits, Norway, for example, has maintained high-quality public services to support informal carers, while in Nova Scotia, Canada, the public sector tends to refuse to provide additional support services to those who receive payments. The Dutch government abolished the attendance allowance that was used to finance both formal and informal services when they experienced an uncontrollable growth of services in the 1970s (Howe 2001). In this policy context, the boundary shifts depend on governments' commitment and strategies and have national significance. The introduction of payment for care appears not always to relieve the burden of family care, but sometimes ties family members to the role unless adequate and sufficient additional support is provided.

In Japan, one of the highly debated points in the planning process was whether or not the new social insurance would reward family carers with cash benefits (Masuda 2002). And the decision not to offer a cash payment to family carers made the Japanese scheme different from the German model. The reasons against rewarding family carers, voiced strongly by feminist scholars and advocates, included: first, concern that a cash allowance would continue to tie women to a caring role; second, it would prevent the expansion of services and providers and thus limit consumer choice if many people opted for cash benefits. Cash has indeed been a preferred option under the German model since 82 per cent of recipients chose cash in 1995, but the rates dropped to 71 per cent in 1997 (Theobald 2003), despite the fact that the cash allowance is lower in value, which is insufficient for people to use it to purchase formal care. The third argument was that in any case formal carers could be expected to provide better care than families. However, this point is questionable because the profile of these newly qualified carers closely resembles conventional family carers without the emotional burden but now with wages. Disengagement of family carers appears to be a far more pressing issue than a rapidly ageing population itself in the development of the Japanese scheme. Some scholars argue that substituting family care is hardly probable under the new scheme because of heavy dependence on informal care in the planning stage, restrictions that make formal care supplemental to informal care in the administration of services; and, finally, family carers being disadvantaged, resulting in a substitution of family care for outsourced services (Morikawa 2001).

Gender issues lie at the heart of the debate. Cash payments encourage continuity of care within the family with some degree of emancipation of women by recognizing their contribution to the family and society through payment. Gender norms and the division of labour are clearly reflected in the construction of German insurance. As Theobald (2003: 174) comments, the possibility to choose between cash and professional packages is meant to encourage care within the family:

> Paid care work within the family, even on a symbolic level, also meets the demand held by certain actors within the feminist movement. Care work should be recognised and be paid for and not be protected/financed by a husband. Women in different life situations, even as carers within the family, could thus become more independent of their husbands and not only have the status as a wife.

In the Japanese context, commodification of care or women's financial independence can be achieved in a different way, which presents a paradox. Sympathy expressed towards the younger generations on the burden of family care provided a strong rationale for the new scheme but it contradicts the reality of 'formalized' carers, since the majority of registered carers are in fact middle-aged women, often married with children, and some of them may be working at the expense of their own family (willing to earn some money by being flexible with their hours). And the types of home help services often provided are domestic work (cooking and cleaning), which many people perform on a daily basis at home without any qualification. The differences in the organization and conceptualization of care in the private and professional contexts should not be underestimated.

Conclusion

Family relations are not static but their profile changes over a lifecourse. Such fluidity of relationships needs to be captured when considering state–family boundaries. Fluidity can also be applicable to the needs for care as people's conditions also change, which requires constant re-evaluation in terms of the cost and feasibility of home care. Once people enter acute periods of illness and decline and hit a point where they can no longer manage at home, a move to institutions may present as a realistic and cost-effective option.

In relation to welfare provision, the boundaries are often drawn by the state through legislation and policy measures, through financing and the provision of certain services or the absence of alternatives. Different boundaries are drawn for different groups of people according to, for example, their gender, family types, ethnicity and income levels. In this context, the expectations that individuals (and families) have over how much the state 'should be paying' or 'should be providing support' are indeed socially constructed.

The logic under which the state and families function is quite different. The state and families do not necessarily perform the same functions or contest over the similar functions. Thus, one does not necessarily substitute the function of the other, and in many societies their functions have shifted with social change. Over the past century, for example, the family's economic needs have been increasingly met by the state through the development of social security. Families now provide more emotional support instead of financial and material support, especially in the national context of developed societies. It may therefore be fair to say that the role of the family to fulfil their own economic needs has been gradually replaced by more importance on providing emotional support.

Whether home-centred care is a low-cost alternative is still a debatable point. In societies where the provision of institutions is low, the provision of home services tends to be equally low. The majority of public expenditure spent on care is often spent on care provided in institutions. This is partly because the majority of those in need who live at home maintain themselves not only through receiving public services but also with the support of informal carers. The cost pressure often comes from various sources, such as a shift in functions from health to long-term care, or an anticipation of cost increase may block effective policy development.

Sustainability is a key issue for those societies that introduced radical reforms to this highly ambiguous policy field, with contested boundaries between the public and the private providers, between health and social care, and between home and institutional care. In Japan, the financial sustainability of the newly introduced social insurance scheme has already become an issue and the significantly shifted boundary has now been pushed back slightly. The review redefined the role of family carers by re-introducing the availability of family resources as an assessment criterion to cut back services. The state–family boundary thus continues to shift according to a nation's socioeconomic and demographic change.

5 Comparative analysis of housing wealth accumulation and family relations

Introduction

Earlier chapters have provided the theoretical and conceptual frameworks exploring intergenerational reciprocity and tried to ascertain why, in contemporary societies, family wealth in the form of housing is important in shaping family relations. The family is at the forefront of a wider support network and it is evident that 'those who have' are often in a strong position to help other family members, to purchase necessary services in the care market, and to build an asset portfolio over generations that would further advantage future generations within the family. Significant cross-national differences have been identified in practices and outcomes regarding asset building and disposal according to gender, culture, housing markets and institutions. Different experiences in generations also exist because each cohort is located in a different socioeconomic and policy context.

In order to gain a greater depth of understanding of the ways in which families accumulate and dispose of their housing assets and to what extent they use (or plan to use) those accumulated assets to negotiate family support from younger generations, a series of in-depth interviews was carried out with two generations in Britain and Japan in 2002/3 and 2005. As discussed in Chapter 1, at the initial stage, the older informants were purposely selected to reflect a range of housing asset levels and personal circumstances. Adult children of the older informants were subsequently approached for interviews to highlight their varying experiences and aspirations regarding access to home ownership, care provision and inheritance. The older informants were all home owners (including a small minority who had already sold their house to move to purpose-built housing for older people). For the adult child informants, the tenure status was flexible and thus mixed. The majority of the British respondents were home owners while the Japanese samples included many public and private tenants, reflecting the tenure patterns of the age groups in the two societies.

Housing wealth and family relations

This chapter will investigate in particular the differentiated processes of housing asset accumulation and aspirations to becoming a home owner between two generations in the two contrasting housing markets of the East and the West. The first section concentrates on older people's routes into home ownership as a part of asset accumulation. The chapter then moves on to explore the meaning of home associated with owner-occupation and to highlight various views and motivations in the shifting housing markets of the two societies. The final part explores the opportunities and constraints shaping the experiences and aspirations of the younger generation accessing home ownership. It is interesting to identify the reasons behind the contrasting patterns that emerged from the interviews with the younger informants cross-nationally.

Names and other details identifying informants quoted in this and the following two chapters have been altered to protect their anonymity. To mark the generations clearly the older informants are referred to by their surnames (for example, Mr Honda and Mrs Duffy) while adult children respondents are referred to by their given names.

Distinctive routes into home ownership

The older generation interviewed in this research were typically born between the late 1910s and the 1920s. Having reached a 'marriageable' age at the end of the Second World War, their adult lives (and thus their housing history since marriage) went hand-in-hand with the nations' postwar development of the home ownership sector in both societies. Prior to 1945, the home ownership rates were less than a third both in the UK and urban Japan (Japanese data is only available for urban areas: Ministry of Health and Welfare, 1941).

In Japan, despite rapid urbanization and the expansion of the number of nuclear households, the ratio of home ownership has been kept at a certain level (around 60 per cent) since the 1960s due to measures used to accelerate housing acquisition (Hirayama 2003b). Since 1950 the government has constantly encouraged the building of owner-occupied houses using low-interest loans provided by the Government Housing Loan Corporation (GHLC). A housing boom is often a reflection of an economic boom. When *Izanagi keiki* [economic boom] hit Japan in 1968, for example, many younger urban families were encouraged to buy their own home. By 1970 the GHLC started offering loans to purchasers of condominiums, which helped shift the profile of the sector away from the dominance of (single-family) houses. Such institutional interventions largely influenced or even manipulated people's perceptions and behaviours towards tenure choice. It was not only public housing policy but also the occupational sector in the mixed economy of welfare that played a significant role in 'mainstreaming' middle-class families as part of creating social cohesion (Hirayama 2003b). The provision of company housing, company loans and saving schemes towards home purchase were some of the examples for companies to assist their employees to achieve home ownership by the time they retired. Such benefits were applicable only to a fortunate few, however, who worked for relatively large companies in full status (see, for example, Hall 1988).

There were roughly two distinctive routes into home ownership for the older Japanese informants: the traditional route through inheritance or by their own means through the labour market participation. Some couples received substantial support from their parents in the process – for example, they were given land on which to build their house. But the most traditional route was a simple one, as in the case of Mr Taguchi (aged 73):

> As the eldest son, Mr Taguchi took over his parental property in which he was born and brought up. The house was old so he had it rebuilt in 1973. He (and his wife) had looked after his elderly mother in co-residency for 10 years before she passed away. He has never lived in rented accommodation or anywhere else but the family residence.

The research by Mizoguchi (2002) on the housing history of factory workers in Hamamatsu City, a medium-sized city in central Japan, also found such distinctive pathways. Almost half of city-born workers in his study had a very simple path to home ownership – they remained in their parental home, married and subsequently inherited the house. If this is considered a male route into home ownership, the female version involved one move from parental home to the husband's home at marriage. By contrast, the housing histories of urban migrants were much more diverse and dynamic. Mizoguchi (ibid.) concludes, however, that a dynamic housing ladder may have been experienced only by those who were born in the certain period associated with high fertility and greater geographical mobility – between 1925 and 1950.

The majority of the older Japanese informants in this research achieved home ownership entirely by their own means, through income reflecting the nature of the industrial city, which attracted a large number of migrants immediately after the Second World War. The city's industrial sector originally developed on the basis of heavy and large-scale industry in the late nineteenth century. In more recent decades, however, the city has not kept up with the nationwide industrial restructuring, and thus its urban growth has lagged behind, resulting in a decrease in population numbers. This has, indeed, had negative impacts on the housing market.

Many older informants started off their married life renting a small apartment (overcrowding appeared to be a real issue in those days), and gradually climbed up the property ladder:

> When I got married in 1959, there were not enough houses because of the war damage. We struggled to find an accommodation and lived in a really

Housing wealth and family relations

small apartment. My married life started in a 4.5 *tatami* room [approximately 2.6m^2] with a wooden fruit box as a table, a pair of rice bowls, two sets of chopsticks and a soup bowl each. We lived there for ten years. Then we were lucky to win a place in public housing – this time we had two rooms [4.5 and 6 *tatami* rooms] plus a small dining kitchen [in total 10 *tsubo* (33m^2)]. There were my mother, my wife and I, two children and another one on the way. It was too small for us. When I opened a newspaper, there was no place to walk around. Then a housing boom came – a construction boom together with doubling real incomes – it was a period we would gain by spending and having a new home built with loans rather than saving. Great inflation! I borrowed as much as I could [from his employer, the National Rail Company] and bought a piece of land. Borrowing was not easy. There were a series of selection criteria and the process took six years. I bought 30 *tsubo* [99m^2] and had a house built, and thought it was my lifetime home! Nowadays, the area has been developed as [a] residential neighbourhood but then it was in the middle of nowhere – a bush behind it, rice field in front, and when it rained we could not walk without boots. After 15–16 years, we got compensation to relocate due to the bypass development. [The current house is a large detached house at the outskirts of the city.]

(Mr Nishi, aged 73)

Such a process of climbing up the housing ladder [*Jūtaku sugoroku*] according to life-course has also been witnessed among the subsequent urban cohorts:

After her marriage, Toshiko moved rented apartments five or six times according to her family needs. When they had their first baby, they used to go to a public bath. But when the second baby was born, they moved to an apartment with a bathroom. When children started school, they moved to a 3LDK *danchi* [public housing] unit. It was small –'… we put three small desks in the 4.5 *tatami* room and put just one futon for three kids to sleep on. Having boys and a girl, we could not do that long time. We both worked [she is a nurse] and thought we could manage'. They bought a second-hand single-family home in the early 1980s, and subsequently had it rebuilt in 2000.

(Toshiko, aged 55)

Climbing up the housing ladder has been an important process for young and growing families, mainly because available rental stock in both public and private sectors was not sufficient in size to accommodate a family or form co-residency, which has been the traditional living arrangement in Japan. Indeed, home ownership was closely associated with the notion of the family in Japan.

In Britain, in contrast, there was limited direct financing by the governments in the 1960s and 1970s to expand home ownership, but instead private banks and building societies were the dominant players. However, the role of the public sector in the direct provision of rental housing has been much more prominent in Britain than in Japan. By 1980 approximately a third of British households were renting in the state system. This was the very reason that helped create a rapid and significant boost to the home ownership sector under Right-to-Buy for social tenants since the 1980s. In contrast, this policy option was not available to the Japanese government owing to the minor role of public housing. As discussed in Chapter 3, the impact of the Right-to-Buy policy is one of the factors contributing recently to sharp contrasts in the growth trajectories of the tenure in Japan and Britain. In Japan, the level of home ownership has been stable, at around 60 per cent for the past two decades, while in Britain it has increased by some 16 percentage points.

Among the older British home owners, the routes into home ownership were somehow similar – the majority achieved home ownership by their own means rather than inheritance or substantial financial support from parents. However, while Japanese households relied more on occupational schemes and the government loans, their British counterparts were assisted by local government in the early part of the postwar period. Few of the respondents had even exercised the early version of the Right-to-Buy from housing associations in the 1960s:

> Mrs Duffy and her husband had been living in the house since 1953 as tenants to begin with and bought it in 1966. '[The authority] used to award points for housing in those days and there was a great shortage of housing after the war and they allocated points on health and service record. Together we [she was ill with tuberculosis and her husband was an ex-serviceman] had quite a lot of points ... We bought it from the housing association. You had an option, they offered you different places. But there was somebody already in here and so we had a temporary prefab and they wanted our prefab so we swapped, we exchanged.'
>
> (Mrs Duffy, aged 80)

For empty nesters, however, the future of their housing and thus assets invested in it is uncertain, especially when the house was located in less desirable neighbourhoods in the city. Over the past few decades those households may have experienced much less capital gain in their property compared with those who have their property in the city's hot spots. This influences their decision on the use and disposal of their housing assets in their old age (see Chapter 6).

A growing family and a lack of suitable alternatives were also some of the reasons for the British respondents to have sought home ownership. In any case, many in this generation had a strong overriding aspiration to become a home owner:

> Mrs Brown started her married life living with her parents-in-law just before the end of the war. Then when they had their first child, the council gave them a prefab, which was still standing there at the time of the interview in 2002.

Housing wealth and family relations

> 'When I fell pregnant again [10 years later] it was only two bedrooms, so the council moved us out. And then gradually, although I had to borrow the money to pay the deposit to come here, I said to my husband, "I want to get out of here". My mum lent me the money to start the mortgage [in the mid-1960s], which I paid her back, and I said to him "I will go and work to pay for the mortgage for the house", which I did ... We could not get back on the council list because I was adequate. And you had to be on the council list for three years. And so therefore the only thing was to try and buy yourself some property, 'cos you were young and you were ambitious.'
> (Mrs Brown, aged 77)

Climbing the housing ladder was also something that the British informants had practised well, including intra-sector upward mobility within the home ownership sector. In both societies, for the older generation, residential ownership was strongly associated with individual houses rather than flats. Home owners in Britain were, however, more frequent movers compared to their Japanese counterparts, whose mobility rates tended to be lower once households moved into home ownership.

> Mr Thompson got married in 1958 and bought their first house in 1962. 'We moved several times after that with my job and bought several other houses. The market was such then that house prices were going up. Every time we sold a house and bought a new one my salary had gone up a bit so we got a slightly higher mortgage until our maximum mortgage was £6,500. It was a big mortgage at the time. And we got a nice house for that ... [When they were separated in the mid-1980s] we sold the house and took £30,000 each, for which you could buy quite a nice little house or flat. We both had jobs and reasonable incomes so we could live independently'. Since then, they have both moved up the housing ladder again a few times and subsequently accumulated substantial housing assets separately in the booming housing market.
> (Mr Thompson, aged 80)

Experience of inheritance

In Britain and Japan, the majority of the older informants did not draw much in the way of financial resources from their parents (or parents-in-law) for their home purchase, nor had they subsequently inherited substantial family wealth. In Britain, Rowlingson and McKay's (2005) research found that almost half of adults (46 per cent) inherited something, but most inheritance involved relatively small amounts – only 5 per cent inherited something worth at least £50,000 (in today's money) during their lifetime. The higher rates of inheritance were found among those aged 60–69, and very low rates among those over 80 and those under 30. This suggests that current older people, who are often the first generation of home owners, missed out from the benefit of trans-generational wealth accumulation, reflecting, for example, the pattern of home ownership (which usually forms the most valuable asset in households).

Similarly in Japan, those who took the self-help route into home ownership tend not to have inherited anything from their parents or parents-in-law. As Mr Soga (83) explained: 'It was very strict in the past. Everything went to my eldest brother. It was the same for my wife's family'.

According to the survey conducted by the Ministry of Internal Affairs and Communications (2003), only 22.3 per cent of household heads with more than two members had inherited something, and the majority (54.3 per cent) inherited a residential property. In addition, the survey conducted by the Tokyo Women's Foundation revealed that 19.5 per cent of the respondents obtained their house through inheritance or a gift (Hirose *et al.* 1998). However, comparing such data cross-nationally needs to be done with caution because the higher percentage of Japanese experiences could be due to the likelihood of household heads being male (who were much more likely to inherit housing assets). On the other hand, gender did not appear to have a statistically significant effect on rates of receiving inheritances of different values in Britain (Rowlingson and McKay 2005).

For the majority who did not receive any substantial inheritance, like Mr Nishi home ownership was closely intertwined with their labour market participation and status:

> 'The house was the biggest purchase in life so I was very happy. But if you think about it, I'd been in debts for a long time because I bought the house with my own means [rather than inheritance]. It went on and on for decades ... In my time, borrowing the maximum amount from the company was enough to build a new house but due to subsequent inflation, the subsequent generations needed [a] company loan plus the government loan. The problem is that you are well tied to the company.'
>
> (Mr Nishi, aged 73)

Home ownership in a changing socioeconomic context

Today access to home ownership has been much more diversified compared with the traditional routes identified earlier, and home ownership may now mean different things to different people cross-nationally. Some examples now follow that highlight the shift from conventional practice.

Traditional route with a modern twist: with demographic change, the traditional route into home ownership in Japan has had a modern twist. The transfer of ownership is no longer restricted to the married eldest sons but possibly to unmarried, co-resident daughters. There are many issues associated with this practice and

Housing wealth and family relations

different conditions (for example, a daughter's financial health, availability and circumstances of other children) produce different views and attitudes among family members (this issue will be revisited in Chapter 7 in relation to the generational contract involving care giving):

> Mari is an only daughter living at home. Her three brothers were all married (some with children) and lived away from home. When Mrs Fujisawa (aged 76) was interviewed in 2002, her father, who had had a series of strokes, was already disabled but was living at home and receiving public services under the newly introduced long-term care insurance. Her parents accumulated substantial wealth including a number of properties, and have already distributed some of their assets to the children. By the time Mari was interviewed in 2005, the ownership of their residential property had been transferred to her. 'About two years ago, while my father was still well, while my father was still able to talk, we transferred the ownership of the house to me because older people aged over 65 can make a gift up to 30 million yen [£150,000] without tax ... Because I have never imagined to marry, and knew I was going to live there forever, because I co-reside, I thought it was mine. The others all left home. I did not want to have any trouble [over the house] once our parents die. So I needed to make it clear that I was going to live there ... I asked for it. Many friends who are not married advised me to do it ... My father has already lost the ability to judge, so I could not have done it now. I did not consult with my brothers, just told them once it was done ... If I decide to sell the house in the future, and the property becomes liquidized, I would divide the money. It is our parents' assets after all. But I need to buy somewhere to live, so would divide whatever was left after the purchase of a condominium'.
>
> (Mari, aged 55)

Home ownership as an investment: in Britain, a market increase pushed housing prices up in many areas and some people benefited from capital gains. This was also the case in urban Japan until the bubble economy burst at the beginning of the 1990s. In terms of housing as an economic or investment consumption good, the specific British characteristic is not only potential capital gains through inflation but also through home improvement. This has been a popular discourse and encouraged by the media (for example, a number of home improvement, property investment programmes on television). This may go beyond the presentation of taste and lifestyle identity in home ownership (Saunders 1990) but may have specific economic calculation. This is a more widely practised approach in a society with a high value retention of older housing stock:

> '[My husband and I] moved to another house we did it up and moved into. We then sold that and moved into another one. Each house was a little bit

Housing wealth and family relations

higher in value ... My ex-husband was very good at doing DIY, he was very good at plumbing and wiring, so we would buy houses that were cheaper because they needed lots of work doing. We would do the work ... We'd live there for a while because it was somewhere we wanted to live and then we'd look for something else. You know so the intention was not we were doing it as a business. It was "Oh, what can we do now?" So we sort of moved up the housing ladder like that, so each house became a bit more valuable'. Caroline subsequently divorced and kept the house as part of the divorce settlement. Now she is looking to trade the house to a smaller inner-city property to meet the needs of herself and her grown-up children.

(Caroline, aged 43)

Condominium ownership: condominium living is a relatively new type of home ownership in Japan. A survey conducted by the Institute for Research on Household Economics (2006) found that virtually none of the condominium owners obtained their current property through inheritance or property transfer. Condominiums are located predominantly in urban areas and it is a little premature to be able to assess the impact of owning condominiums on trans-generational asset accumulation:

Hideo and his wife bought a second-hand condominium when they got married in 2000 in suburban Tokyo. 'One of the reasons is that it is economically viable to pay a housing loan than a rent. In Japan you can find better quality condominium for the same price for rental properties. If we rented this property, it would cost us ¥200,000 [£1,000] per month, and we cannot afford it. It gives us a status but above all it expanded our freedom. It is my dream to build our own single-family home but I place higher on use values than on potential capital gains. It is important to separate the two values. I prefer to live in a condominium [better location and more affordable]. I am not attached to a piece of land as the previous generations may have been.'

(Hideo, aged 35)

In the recent climate of housing market stagnation, people do not necessarily expect home ownership to be the best way to achieve wealth accumulation through capital gains. Hideo, for example, thought use values (the property to achieve a better quality of life) more important than owning a piece of land that was traditionally believed to hold value higher and for longer. As Figure 5.1 indicates, over the past three decades in Japan, the attitudes of owners towards condominium living have shifted significantly. A marked shift came in the early 1990s when the bubble economy burst. The lack of supply of alternative high-quality rental accommodation is one of the primary reasons behind the continuous popularity of condominium purchases despite its declining values as

Housing wealth and family relations

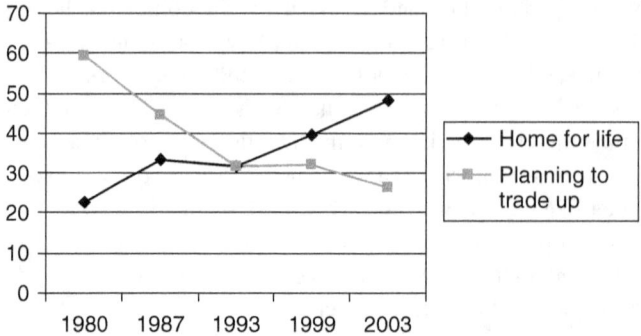

5.1 Attitudes of condominium owners in Japan.
Source: Ministry of Land Infrastructure and Transport (2003).

a tradable commodity, and less of a likelihood of asset accumulation compared with single-family homes (Yoneyama 2005). Condominiums are thus no longer necessarily considered as a stepping-stone to single-family homes.

The meaning of home in the East and the West

Home ownership plays similar as well as different roles in Britain and Japan. A substantial body of literature has theorized the meaning of home and home ownership, particularly in the Western context (for example, Ronald, 2008; Gurney 1990; Saunders 1990; Dupuis and Thorns 1996). For example, Munro and Leather (2000: 514) argue that 'the economic distinction between housing as an investment and consumption good has parallels with sociological discussions of the meaning of home associated with strong ideological norms and values'. In the policy context owning a home may be associated with self-identity, cultural expectation, an investment, or simply an alternative solution. The meaning of home can influence how people make particular tenure choices, and also how people treat their housing assets in their old age. While this section does not expand the debate on the meaning of home, whether as ontological security or pride of possession (for detailed debates see, for example, Gurney 1990), rather it highlights significant areas of similarities and differences between the two societies. The meaning of home will be revisited in Chapter 6, in relation to people's plans for bequests or using their housing assets in later life.

In Japan, a hypothesis used in Hirose *et al.*'s study (1998) states that housing is often associated with the family, for example, family continuity – housing assets do not belong to the individual but to the family and are thus passed on from one generation to the next. This legacy of 'family continuity', or keeping the family line going, is indeed deeply rooted in the mind of some older people, often

Housing wealth and family relations

accompanied by some ritual traditions. Mrs Morita felt strongly about her responsibility of maintaining 'ancestral assets' for future generations. Being the eldest child (she has a younger sister) in the family, she received a *yoshi* (adopted) husband and succeeded the family surname and properties. This practice was perpetuated to the marriage of her only daughter. Her daughter initially took her husband's surname at marriage but before their first child went to school, Mrs Morita adopted back both her daughter and son-in-law to keep their family line going:

> 'It was my father's wish. My sister does not have a child so he did not want to let the name disappear. There was an issue of family grave as well … I cannot increase our assets anymore but try not to lose any. I am telling my daughter that she could sell some properties if necessary – liquidizing some land currently used for a parking lot to cover inheritance tax, for example. But I do not want to be the one who's sold our family wealth. I did not buy those properties myself – these were passed on by the previous generations!'
>
> (Mrs Morita, aged 74)

The importance of family continuity for the majority of respondents was, however, contradicted by their family and employment background. The majority had been employed (rather than engaged in farming or the family business), did not inherit anything (or if they did, it was not very much) from their parents, and thus access to home ownership was achieved entirely by their own means. Hirose and colleagues' (1998) survey also confirmed the strong willingness of the first generation of home owners to leave housing assets to their children.

Home ownership representing stability appears to be universal. For Richards (1990), 'security' extends to 'security for the future' in relation to family unity and stability and also to meanings of settling down, foundation and permanence. Home ownership in Britain was also strongly associated with the family:

> Jackie and her husband first had a flat for a year or so and then got the house in which they have lived for 21 years. 'We just wanted to start a family so [owning a house] is just a natural thing to do I think. It is something that you leave your children, isn't it.'
>
> (Jackie, aged 46)

Another factor related to this point was the attachment to their own land and dwelling. In Japan, once people owned a single-family home their mobility tended to be low and trading up was often achieved by rebuilding a home on the same site. Therefore, some people were attached to the *house*, not just to the *assets* invested in the house. However, this adds to the contradiction. In many ways, the aspiration of home ownership for this generation was institutionally driven as owner-occupation in the postwar period symbolized security and status more than a mere investment. The pride of possession may have thus overridden the functions of housing for use or as an investment. In the current economic climate, a survey on women and housing assets revealed that the majority of the respondents (56 per cent) mentioned security of tenure as a reason for owner-occupation and virtually none (1.2 per cent) mentioned capital gain expectation as the reason (Institute for Research on Household Economics 2006). But the survey respondents still make an economic decision (53 per cent) – owner-occupation is better than paying rent considering that with the same money one could only get lower quality and a smaller space in the rental sector.

The younger generation in Japan had different views on the sentiment or attachment that their older counterparts expressed towards home ownership. Some of the comments suggest that they associated housing more with family relations – experiences of living with their parents and other members – than the physical structure or invested wealth. Their view can thus be more pragmatic:

> [When our parents pass away] I think we have to sell it and divide the money. I am not attached to the house. We [children] have all left the home – the house was important when the family was there. It is important that our mother and father live there – the house itself is not important. We all got new family and live our own lives, that's part of social change and have to accept it.
>
> (Masami, aged 45)

A similar view was expressed by some of the British informants. For example, Caroline was much more attached to her parental home than her own house because of the family relations that had provided stability and the foundation of her life:

> I have more attachment to this [parental] house. Yeah, you know if they wanted to move I'd be quite happy to buy this from them. I haven't told them that, but I would. Because it is ... it is the family house ... I have far more attachment to this house than I do to any of the other houses I've lived in ... it was a lovely house to grow up in. And it is home. This is coming home for me ... I mean when my husband and I split up I moved back for a month or so with them while we were sorting it out, and it is home ... because I see [my current house] very much as an intermediate solution, I always have done. It is practical. The reason that I bought that specific house was'cos it suited the family needs, not because I wanted to live in that particular house, which is a bit different.
>
> (Caroline, aged 43)

Housing wealth and family relations

In Britain, access to home ownership has become very much a normative expectation. Owning a house has become a middle-class 'expectation' and the second generation of home owners, as the following quotes suggest, do not even think of any other alternatives:

> It is security, it is not investment. Well it was not seen as a monetary investment it was seen as security. Your own place to live in. And sort of a middle class and professional expectation. Professional classes expected to own their own home.
> (Paul, aged 63)

> I think it did not even cross our minds not to buy because it just seems that as soon as you can, you get on the housing ladder, otherwise you sort of miss it. You know and the money that you are paying towards renting can go towards a mortgage and it is for your future, you know.
> (Ellie, aged 46)

In both societies, however, home ownership is no longer only a label for middle-class families. Membership has now been extended to include a wider section of society such as low-income households in Britain and also single-person households. In other words, the latter suggests ownership of housing is now less closely associated with family formation.

Economic incentive as an expectation of capital gains through home improvement or market increase may not be the primary factor to own a house, but people often weigh up the economic value for owning as opposed to renting. Moreover, some people consider home ownership as an accumulation of assets that can be liquidized in their own old age. Such an asset-based welfare idea has been more progressed and widely accepted in British society compared with Japan, where there are more barriers, both cultural and institutional, for turning housing stock into cash flow (Izuhara 2007):

> I think this is more the culture here I have to say. I think for me, my home and my environment is very important and probably the security of the home and environment's important. Whether it is actually important to own it as opposed to rent it I would not say necessarily except we have much more that sort of culture here in Britain anyway. It is important, you know, you get on the property ladder ... And also now I mean that was encouraged right from the start. But now that there is this big sort of scandal over pensions and the fact that people are not ... we realised the property we have in a London suburb, we are very lucky, we got it at a good price and we did a lot of work to it. Essentially that is our pension now.
> (Penny, aged 54)

Women and accumulation of assets

The cross-national research highlighted various gender aspects in routes into home ownership and the process of housing wealth accumulation. In Japan, where the family and labour market were largely organized around the male-breadwinner family model throughout the postwar period, home ownership was often achieved entirely by male incomes. Despite an increase in the labour market participation rate for women, there remains a dip in participation rates at ages 25–29, continuing through to ages 30–34 as women exit the labour market to care for children and other family members (Atoh *et al.* 2004). Male and female participation rates are about the same until the age of 25, but by the age of 30–34 the women's rate is around a half that of males. This pattern of women's labour market participation is not evident in the UK. In 2001 57 per cent of mothers with children under five and 74 per cent of those with children between five and ten were in paid employment (although the majority were in part-time work) (Women's Equality Unit 2002). Those rates have increased steadily throughout the postwar period. In Japan, therefore, in the period of postwar great economic growth, those households who had substantial contributions by women helped advance their assets status further:

> After coming back from Mainland China, we got married in 1947. There was not any job in our home town so we moved to this city. We borrowed ¥100,000 [£500] from my husband's brother and bought a second-hand house in 1952. Ten years later, we traded it up to a new built-for-sale house with a loan from his company. We then bought a large piece of land, and built a large detached house and a dormitory for bachelor employees and I earned my income preparing meals for them. I was earning more than an average salaried man. Otherwise, we could not afford having such a large house. [They subsequently bought a number of land and properties in the area.]
>
> (Mrs Fujisawa, aged 76)

> Mrs Suzuki was a civil servant until she retired at the age of 50. Her mother saved all her salaries and gave it back to her as dowry at her marriage. Her mother's support with childcare helped her continue her career with a respectable salary. Since they were living in an owner-occupied house in rented land, they bought a piece of land near-by with the intention of building their home there. They subsequently bought the land that their house sat on from the landlord, resulting in them owning an additional property (which was eventually divided into two for their two children to build a house each).
>
> (Mrs Suzuki, aged 71)

Some double-income couples of this cohort successfully boosted their asset profile well above average, which was effective given the timing with postwar economic growth. However, this was not always the case. Other women like Mrs Brown (aged 77) also worked but only supplemented their husband's income to achieve a single, modest owner-occupied property.

Another interesting cross-national difference is how women's financial contribution has been translated into their ownership status. A survey on women and material assets in Britain and Japan revealed Japanese women's disadvantaged position in asset holding (Izuhara and Kennett 2006). In Japan, ownership by the husband alone was the most common pattern of ownership across all age categories (73 per cent of the samples) while the majority of the British sample with a mortgage were either joint (76 per cent) or sole (21 per cent) mortgage borrowers on their property. With social changes, however, younger women in contemporary societies appear to be more assertive in staking their claim in relation to housing. Hideo (35) and his wife (who works full time) own their condominium jointly:

> Hideo: 'The reason why we decided to own jointly is something to do with tax concession. Well ... why did we decide to be joint owners?'
>
> Wife: 'That is because we are both contributing towards the mortgage'.
>
> Hideo: 'That is right. It's not because of taxation, but because we are both paying for it'.

Divorce and separation often disadvantage women in their housing situation and accompanying asset accumulation. However, the research found some cases in which divorce and separation created an opportunity for women for asset accumulation. Less common arrangements (a parent–child partnership) were also found as part of this demographic shift:

> Michiko had moved apartment every five years since her marriage. Right after they bought their first condominium she got divorced from her husband. She was 32 and went back to a square one renting a small apartment with her two young children. She bought a condominium on her own 10 years ago. 'Thirty-five-year loan but monthly repayment was cheaper than my rent although I need to pay a lump sum twice a year at the bonus seasons. If you think a lot you cannot make such a big purchase. I decided it straight away. A 35-year loan means I would be over 80. It was an issue and also it was difficult to buy as a woman on her own – needed a guarantor. I borrowed the money from the Government Housing Loan Corporation. It costs me a lot not just mortgage but also property tax, interest rates which will go up after the fixed period, necessary repair, and so on'.
>
> (Michiko, aged 55)

Housing wealth and family relations

> After Mrs McKenzie's divorce, she and daughter Jennie (53) provided each other with financial and other support, and always bought their houses jointly partly due to practicality – they needed two incomes to be able to purchase a property in London. Mother and daughter became partners to replace the husband. However, there have been dilemmas in being co-owners in terms of dependency and differences in lifestyles. The mother and daughter finally sold the house and split the money when Mrs McKenzie moved to sheltered accommodation.
>
> (Mrs McKenzie, aged 82)

> When Mr and Mrs Thompson separated 20 years ago they split the money from the sale of their house and each started accumulating their housing assets again separately. They were both teachers and had a reasonable income. Mrs Thompson died with cancer right before I interviewed their daughter, Pam (aged 43). In her will, her assets including a property worth £220,000 would be split three ways to her husband and two children. If they had stayed together, they could have had only one property, and inheritance would not have skipped the widowed father. 'It is the fact that if mum and dad were together in the same house we would not now be inheriting obviously mum's money ... it is only simply because they had separate properties that this is happening ... The world is open with many opportunities. We can do anything and [husband] do not have to do a pressured job. It is no good pretending that money does not matter'. The good timing meant that Pam and her brother Ben (aged 46) could improve their housing situations. Pam will pay off her mortgage. With inherited money Ben, with his new partner, can now add one more room or find a house in a better location in London.
>
> (Mr and Mrs Thompson, aged 80 and 69 respectively)

The impact of inheritance on the younger generation will be explored in more detail in Chapter 6.

Contrasting aspirations among adult children

Having been property-owning societies for decades, there has been a consensus among the older generation on the importance and value of home ownership within and across the two societies. Contrasting housing market trajectories in more recent years meant, however, that such a consensus has become absent among the younger informants, especially due to the stagnated housing markets in Japan. In the period between 1997 and 2005, for example, house price indices increased by 154 per cent in Britain while Japanese indices declined (–28 per cent) (*The Economist*, 16 June 2005: 74). It is easy to assume that housing market conditions largely influence people's perceptions on home ownership as a tenure choice. According to Hirayama (2006), for example, even between 'baby-boomers' (born between 1946 and 1950) and 'baby-busters' (born between 1956 and 1960), there are significantly different views expressed on home ownership owing to their timing of entry, the experiences of capital gain/loss and economic recession. In Britain, in contrast, even among households in negative equity in the mid-1990s, Forrest and his colleagues (1999) found a substantial commitment to the home ownership ladder and a strong belief that a mortgage was buying something in a way that rental payments were not. In Japan, such belief appears to be no longer shared by everyone.

Comparing home ownership rates by age reveals some interesting differences (see Figure 5.2). In Japan, there is a clear correlation between age and housing tenure – the older one gets, the more likely one is to own a house. People's housing career is often developed in parallel with their family and labour market career. There is thus a marked increase in the level of home ownership beyond the age of 35, indicating the generally later entry of households into home ownership compared with Britain. Over the past two decades, there has been a particular decline in home ownership rates among younger cohorts (from 9.9 per cent in 1978 to 3.3 per cent in 1998 among the under-25 age group and from 27.9 per cent to 12.7 per cent among those aged 25–29), partly due to growing income and employment insecurity with the prolonged recession (Forrest *et al.* 1999). In times of economic instability, family welfare tends to absorb potential new households. According to the 1995 Census in Japan, approximately ten million single people aged between 20 and 34 were still living with their parents. The issue regarding 'parasite singles' had become a popular discourse at the turn of the new millennium, as discussed in Chapter 2 (Yamada 1999). However, subsequent research attempted to construct a more holistic picture of this cohort, highlighting their contribution to the household economy and complex family relations. Although in Britain the 25–34 age cohort is generally considered to be first-time buyers, the similar phenomena of stagnated entry to the sector and return to parental home has also been witnessed, and there was a reduction in owner-occupation in the younger age groups in the late 1990s (Council of Mortgage Lenders 2001).

The difference in the timing of entry may be explained by the different roles, functions and meanings that home ownership plays in each society. Although such practices and meanings are often multifaceted, in a society where home ownership is considered predominantly as an investment value and asset accumulation, this tenure is preferred regardless of age and marital status. However, where the significance of home ownership is associated more with family relations, marital status becomes a strong determinant. Thus, not only the precariousness of employment and income, but also declining rates of marriage among those in their late twenties may be partly a factor to explain the declining rates of home ownership in younger age groups in Japan. The promotion of 'independent

Housing wealth and family relations

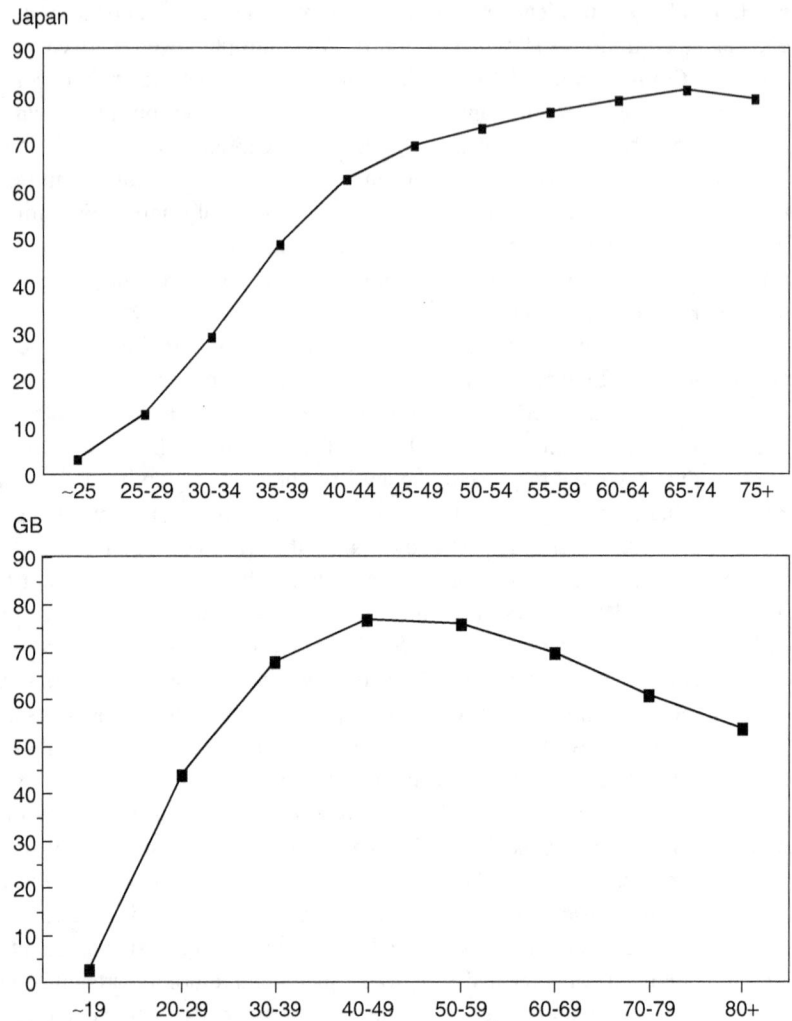

5.2 Home ownership rates by age of head of household, 1998.
Source: Japan: Statistics Bureau, 1998 Housing and Land Survey of Japan; GB: General Household Survey.

living' as opposed to 'co-residency' is another cultural (and also economic) factor which influences household formation and consequent tenure patterns. Young adults tend to leave their parental home earlier in Britain regardless of family formation, which results in more opportunity and motivation to enter the sector early. However, the stagnated entry has also been the case due to the more recent housing boom, especially those living in hot spots. (In Britain, the average house price hit £200,000 in 2007, which that means that single earners find it difficult to enter the sector.) All the British adult-children respondents in the research were home owners, and the concern was mainly felt towards the housing futures of the next generations:

> Paul's daughter was renting at the age of 35. 'I can't see any way she can get into the housing market unless she has an inheritance. But even then she couldn't afford the mortgage on her sort of income. Her income is about £6,000 a year [being a musician]. It is not stable. She is on Housing Benefit'.

> Parents often do when their children show some signs of either being too independent they wish weren't, or if they are not independent they wish they were. And we feel a little bit you know it is a shame that our daughter [aged 24] has not forged her own path already, you know that she's got groups of either a boyfriend, or a partner or just friends that she wants to live away from home. And so we feel a little bit that it is a shame that she is back again from that point of view.
>
> (Penny, aged 54)

In Japan, on the other hand, such concern was directed to many of the adult-children respondents themselves. And it was not just that their willingness to own was compromised by the housing market volatility since in a way houses have become more affordable in the current market, especially second-hand properties. But people's perceptions towards housing tenure appear to have shifted. In the current economic context, the younger cohorts appear to be less confident in owning a house as a vehicle of providing security and stability. Home ownership may no longer be their primary goal and most desirable housing tenure. Some have taken more pragmatic decisions in their tenure choice. While the British respondents had consensus over the importance and value of home ownership, their Japanese counterparts expressed more ambivalence and diverse views, with competing aspirations. Out of 16 Japanese adult-children respondents, seven informants were either renting in the private sector or still living with their parents (as opposed to every informant being a home owner in Britain). There were both opportunities and constraints influencing the tenure choice of the respondents and there are several points to highlight such changing aspirations towards home ownership among the interview cohort.

In Japan, the circumstances of home ownership are more complex, with the values and sometimes ownership status of housing and land often separated, and the longevity and durability of housing (the building itself) adds a further complication in how the use values are weighed against various other factors.

Housing wealth and family relations

Timing matters to some respondents (in terms of the age of housing stock), who felt it daunting to face major repairs or even rebuild in their old age:

> Both Chiyo and her husband are teachers and have been tenants throughout their married life. They have no children. They are planning to have a new house built when she retires at the age of 60. 'Japanese houses become shaky after 20 years – such as floors wobble. That is why we do not want to buy an old [second-hand] house. I will be 80 in 20 years time. That is why I would like to have a new one built at 60 and make it last for the rest of my life [without major repairs]'.
>
> (Chiyo, aged 56)

Some informants opted for renting for various reasons in the current socioeconomic and labour market context. The following comments capture well the multifarious nature of such motivations:

> Yuki lives with her husband and two boys (aged 7 and 5) in rented accommodation [condominium]. 'The primary reason why we are renting is that we would like to move according to our life-course needs. Our last move was because we wanted to be in the catchment area of our children's school – it's more like an after-school club which ethos I greatly agree with. And the second reason is that I do not want to be in large debt. Perhaps that is more important. If we borrow 10 to 30 million yen [£50,000–£150,000] I cannot quit my job [as a full-time nurse] even if I wanted to. That puts me off. For example, I am working full time now, like mad, and if I want to stop doing it, I would not have the choice if I had a huge debt. I know many people around me who are tied to their job because of their housing loan. There is another reason – can I tell you? My father-in-law has residential properties and we got to do something about them in the future, whether we like it or not. If we bought our own house, he would be really angry. He is expecting we will rebuild his old house and live together'.
>
> (Yuki, aged 41)

In considering the nature of housing as an exchange commodity tenure flexibility is something distinctively Japanese. In the British housing context, owning means less constraints in people's flexibility in terms of trading their properties up or down for various reasons (although there were 'frustrated movers' in times of negative equity) (Forrest *et al.* 1999).

> Hajime and his family are renting a 3LDK *danchi* unit. He expressed very complex and ambivalent feelings towards his housing destination as the eldest son of Mr Taguchi – traditional duties as the eldest son, relationship

with his wife's widowed mother and his financial status being a freelance writer. 'The reasons why we are renting this place are primarily location and neighbourhood. Safety for children and having plenty of green space are appealing factors. It has a good transportation network and also is located roughly in between my family and my wife's ... In fact we tried to buy a condominium last year but the bank refused to lend me money. That is because I am self-employed. And it is also because my father owns a property my relatives were against my intention. No one said anything out loud but I felt lots of unspoken pressure'.

(Hajime, aged 45)

Some research interpreted the stagnated entry to home ownership by the younger generation as a greater potential for inheritance to weaken their aspirations to get on to the housing ladder early. In the survey conducted by the Ministry of Land, Infrastructure and Transport (2003), for example, 27.3 per cent of households mentioned the possibility of housing inheritance. This may be especially the case in the context of the 'capital loss market' – early entry to home ownership does not create any advantageous conditions, and it can rather be disadvantageous due to the short lifespan of housing. However, the research found a more complex relationship between tenure choice and family relations beyond this mere 'stand-by' characteristic. Some felt that the traditional duties (of the eldest son) and property inheritance were more of a burden than a welcome opportunity:

We do not want either their house or land. We would prefer they [husband's parents] sell it and live wherever they want. But they do not do it ... When I had our first baby, they gave up everything they were doing such as travelling and have become 'available' for grandchildren. One of the reasons why I was able to continue my career is their support on child care. But for them, looking after grandchildren equals living together, and I am sure that they are expecting for their favour to be returned in the future. I do not want money! I do not want to live with them.

(Yuki, aged 41)

Succeeding his parental home also means co-residency for Hajime's case, but he was also very ambivalent about going back. Partly he was not confident about his financial ability to have the father's house (built in the early 1970s) rebuilt when the time came.

(Hajime, aged 45)

To avoid becoming responsible for their parent(-in-law)'s property or to avoid co-residence (which is anticipated as more problematic than anything else, some adult children make a (*de facto*) statement by purchasing their own house, thus avoiding succeeding to their parental house. In this scenario, the route to home ownership is more by the avoidance of traditional practice.

Conclusion

The cross-national research highlighted that in both societies the older cohort was largely the first generation of home owners in the family, who started accumulating housing assets almost entirely by their own means through labour market participation rather than asset transfer across generations. This process was assisted by the postwar development of home ownership as well as economic and real income growth in both Britain and Japan, although the balance and types of assistance between the state, market and corporations, and the target of the assistance varied. Among the older generation cross-nationally, there was a strong consensus on the value of home ownership both within and across the societies. There was no question in this cohort that home ownership was the most desirable tenure and 'everyone aspired to it'.

When looking at the cohort of their adult children, in contrast, a different picture has emerged. The British informants continued to exhibit a strong brief on the value of home ownership, while the Japanese views were much more mixed and ambivalent. The contrasting trajectories of the housing markets for the past two decades may be one of the explanations behind the cross-national difference. Family relations and practice could be another.

Among the British adult-child informants, all of whom were home owners, the concern over access to home ownership was particularly addressed not towards themselves but towards future generations. With housing price inflation, affordability has indeed become a major issue for younger people in Britain in more recent years, and many schemes (for example, for key workers in London and part-ownership) have been introduced to help access to home ownership. In Japan, where the housing market has stagnated since the burst of the economic bubble in the early 1990s, the views expressed by the younger generation were diverse and ambivalent. For example, some people positively planned to remain as tenants throughout their life-course, and others saw condominiums as their housing for life rather than a stepping-stone to single-family home ownership. For some, housing assets owned by their older parents were seen as a burden rather than an opportunity. Indeed, the Japanese fieldwork found the persistence of cultural practice in relation to housing assets and family relations that highlighted as many constraints as opportunities in their housing and lifestyle choices.

6 A comparative analysis of perspectives on inheritance

Introduction

Chapter 5 drew on a series of in-depth interviews with two generations in Britain and Japan to explore their access to and experiences of home ownership as well as their aspirations to become home owners. It also examined the extent to which the family was behind the motives for accumulating housing wealth. Having established the processes of housing asset accumulation, this chapter focuses on the other end of the life-course spectrum – examining people's views and practices regarding the disposal of their accumulated assets over the generations. The comparative analysis will again reveal how much it has cultural resonance using the voices of two generations in the East and the West.

The chapter begins with the classic questions around how older parents plan to pass on their wealth and to whom. Cultural ideologies associated with inheritance run strongly in the views of the informants in this area and are used as a tool to examine the cross-national data. The chapter then moves on to discuss how current socio-demographic changes influence their decision-making process, or more precisely, how changes in an individual's life-course affect their perceptions and plans of bequests reflecting both potential and actual changes (that occurred in the course of the research). With profound social change in more recent years, combined with the ageing process of individuals, study of inheritance needs to capture such fluidity in various aspects of human life. The impacts of inheritance on the younger generation and alternative strategies of using housing assets in old age are also analysed in this chapter. Overall, the key question is whether inheritance, or intergenerational transfer of family wealth, reinforces solidarity or causes conflict among family members.

How to divide housing assets

Equal share versus one-son succession

> [My assets] are going to my two children. [Interviewer: 'Equally?'] Yes … Well, they've both got their own homes, but they are equal to me … They are equal. They are very different but both equal.
>
> (Mrs Davies, aged 70)

Analysis of perspectives on inheritance

> Even if I were not living with my eldest son, I would leave this house to him. You may ask why – it is not because I like him the best, but it is because he is the successor of the family. The eldest son will take in both good things and bad – everything including the family tomb.
>
> (Mrs Tanaka, aged 86)

The above contrasting responses represent stereotypical ideas regarding property transfer in the two societies. In general, British data on attitudes and practices towards inheritance and family support were more predictable and exhibited a greater degree of consensus within and across the generations, while Japanese respondents were more divided, highlighting the process of social transition such as value shifts (Izuhara 2004). Low fertility also contributes to the change in Japanese practice.

Following the civil code issued in the postwar period, although some older Japanese informants were planning to divide their assets equally among their children like the majority of British respondents, there were still a considerable number of people such as Mrs Tanaka who wished to leave their housing assets intact to one particular child. In Japan, 12 out of 17 older owner-occupied informants who had more than one child were planning to leave their house to one particular child, compared with only two informants in Britain (14 older informants were planning to divide the assets equally among their children; four informants had one child only; and another four had already sold their house to move into supported housing). There are a number of socio-economic and cultural factors underpinning such cross-national differences, including the different roles home ownership plays in each society.

Considering housing as family assets (instead of the assets belonging to one particular individual or couple) is one of the strong reasons for the conventional practice in Japan (see Chapter 5). Such a symbolic 'meaning of home' sometimes makes it difficult to divide assets invested in housing. This legacy of 'family continuity' and ancestor worship is indeed deeply rooted in the mind of some older people, often accompanied by ritual traditions:

> I would like to leave it to whoever will look after it – to the person who will look after the ancestor and the Buddhist altar that is the primary condition. And then the person can also look after us. It does not matter whether son or daughter. [Interviewer: 'Not thinking about dividing it equally?'] No, because you need money to succeed the family, carrying out ritual ceremonies, maintaining the family tomb and so on.
>
> (Mr Akiyama, aged 76)

In the Japanese context, passing assets intact to a successor child thus provided some families with the foundation of continuity and maintaining family solidarity.

Analysis of perspectives on inheritance

It is not surprising therefore to witness such arrangements in families in traditional co-residency or those engaged in farming or a family business. However, the importance of family continuity was contradicted by their own inheritance experience. The majority of the older informants had been employed (rather than engaged in farming or a family business), they did not inherit housing assets from their parents, and thus access to home ownership was achieved entirely by their own means. The strong willingness of the first generation of home owners to leave housing assets to their children was also confirmed in other research. According to a survey conducted by the Ministry of Telecommunication in 1998, it was not only those who obtained their housing through inheritance (72 per cent) but also those who bought housing by their own means (58 per cent) who wanted to leave their house to their children, and even among tenants, the largest proportion (38 per cent) were willing to leave their savings to children. In other words, the importance of the family was certainly behind their motives, but inheritance is not necessarily part of a preconceived idea of the chain of obligations across generations. The older generation were often caught between the traditional (pre-industrial) customs and the modernized sociolegal systems that inevitably produced contradictory and ambivalent responses.

Housing as commodity?

When the question of bequests was posed, treating a home as an itemized individual unit or part of one's overall assets presented a differentiated 'meaning of home' cross-nationally. In Japan, housing (land) was often treated as a separate unit from other financial and mobile assets, and a combination of use value and symbolic meaning of ownership tended to produce some specific responses:

> Mrs Suzuki gave a plot of land to her son and daughter who built a house next to each other. Her daughter has no child, but her son has three. 'Both of my children have their own house but I have three grandsons. So [with mine included] the boys can have a house each'.
>
> (Mrs Suzuki, aged 71)

According to the survey conducted by the Cabinet Office in Japan (2001), there was a clear urban–rural difference regarding the whole property transfer. Only 39 per cent of those living in large cities think 'residential property should be passed on to children intact' as opposed to 72 per cent of those in small towns and villages and 67 per cent in local cities. And the older people are, the more likely they think in this way. This explains why such tradition remains stronger in the minds of older people in rural communities.

In Britain, by contrast, housing has been considered more as a 'commodity' and treated as part of one's overall assets. The commodification of housing has

Analysis of perspectives on inheritance

been accentuated through the postwar development of the owner-occupied sector (Forrest and Williams 1984). British people, in particular among the middle class, are more frequent movers, and beyond a very small minority there is little attachment to a particular piece of land. However, this point differs from the debate on people's attachment and the meaning of the home in *later life* – more specific meaning is often attached to housing and a living environment (Heywood *et al.* 2002). In Britain, as Finch *et al.*'s (1996) study on wills indicated, residential property was often treated as part of a person's overall assets, and older parents expected their children to sell such property and divide the monies equally among themselves. Even in the rare cases of co-residency, housing is still a commodity, and the expectation of equal treatment can supersede the housing need of a co-resident child. Mrs Porter had two daughters and was living with her unmarried daughter at the time of the interview. She was sympathetic to her daughter's housing needs but her will was still based on the principle of equal share:

> What worried me is if Susie's living with me and I pop my clogs I do not want the house sold to be divided up between the two. So I have stipulated that it is got to be at least six months before the house is sold so that Susie has time to look around. Those things you have to look at, isn't it? Whereas my mother-in-law, when she lost her father, she had to get out instantly for the house to be sold up and divided between her sisters. Well, I do not want that to happen to my two daughters, you see.
>
> (Mrs Porter, aged 71)

In another co-residency case, Mrs Gaskin sold her house and put half of the money raised from the sale to her current five-bedroom house that she shared with her daughter's family:

> It was the only way that her daughter and husband (with three grown-up children) could afford such a big house in their neighbourhood. And also because she was widowed so young, her pensions were low and it made more economic sense to live together. The arrangement works well – she has a room out onto a patio and a large bedroom with an en suite bathroom above. It is like a maisonette, having her own quarters and her privacy is well respected. She was keeping the other half in reserve for her other daughter. Their jointly purchased house doubled its value within six years, but her saving is 'hardly getting anything on it'. However, Mrs Gaskin holds onto the saving because 'I feel I would like to give her an advantage as the other one has had'.
>
> (Mrs Gaskin, aged 78)

In Japan, on the other hand, *dōkyo* (co-residency) is a more shared arrangement in terms of space, contact and resources. Postwar trends have seen a dramatic

Analysis of perspectives on inheritance

shift towards independent living, which has resulted in an increase in the number of elderly-only households, while 'three-generation' extended families continue to decline (see Figure 6.1). Another interesting trend has been an increase in elderly households living with unmarried adult children, reflecting recent marriage patterns in Japan. In Britain independent living is a conventional living arrangement and single-person households are common; according to the General Household Survey in 2005, 19 per cent of men and 33 per cent of women aged between 65 and 74 lived alone and 29 per cent and 60 per cent respectively for those aged 75 and over did so. In Japan, among older people, the older age group (85+) is more likely to co-reside (71 per cent in 1990) than their younger counterparts (53 per cent among those aged between 65 and 69) (Ministry of Health and Welfare 2000). Although some argue that it is a temporary arrangement or a mere postponement until their circumstances change (Hashimoto 1993), a steady decline in extended family living among all age groups may require a different explanation.

Moreover, the composition of *dōkyo* families has moved away from the traditional three-generation household where older people live with the eldest son and his family. This is partly linked to the changes described in family formation. For example, despite the general decline in co-residency, the number of older people living with unmarried children has increased from 16.5 per cent in 1980

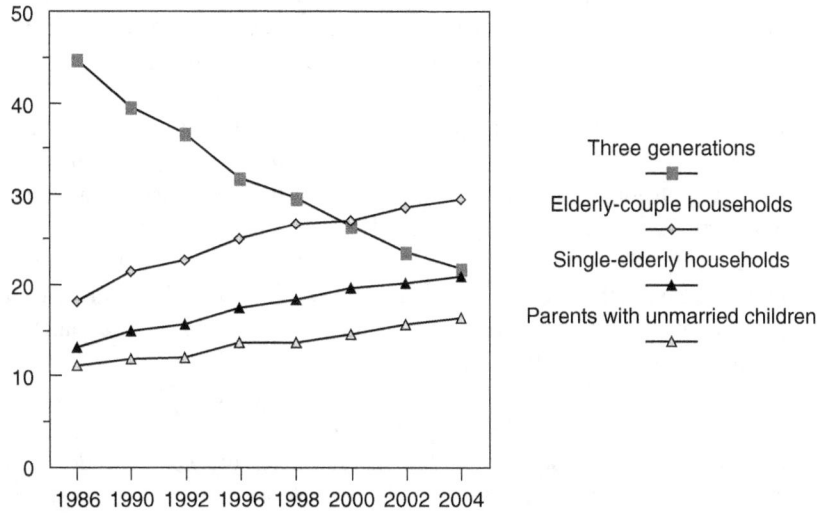

6.1 Type of households among those aged 65 and over in Japan.
Source: Ministry of Health and Welfare, Kokumin Seikatsu Kiso Chosa (1986–2004).

Analysis of perspectives on inheritance

to 19 per cent in 1998, which forms 38 per cent of all co-resident households (Ministry of Health and Welfare 2000). Unlike the English tradition of keeping the youngest daughter at home as a carer in the past, these unmarried children are likely to be in employment. Living with the family of one's own daughter is also increasingly popular in Japan. Living nearby without sharing accommodation is another effective way of maintaining both independence and family support. The ratio of older people living near to their children (either as a separate unit in the same house, a separate house on the same land or in the same neighbourhood) has increased from 9 per cent in 1986 to 12.7 per cent in 1998 (Ministry of Health and Welfare 2000). Such family nuclearization is likely to complicate care responsibilities pertaining to the family that had been the mainstay of existing social policy, but at the same time such physical separation may result in a better relationship between carers and the cared-for without involving a considerable emotional burden. Indeed, it is such changing living arrangements that have contributed to alter conventional inheritance practice in Japan.

Inheritance and family solidarity

Returning to bequests, Mrs Matsuo was planning to leave her newly built house to one particular son and expected the house to remain in the family and to act as a magnet for family gatherings after her death. Her husband deliberately willed it to their second son who was financially and socially stable compared to their eldest son. Mrs Matsuo did not mind which child moved in to live in the house as long as the house stayed in the family. In her mind, leaving the house (real estate) instead of financial assets (retirement allowance and savings) was the key to keeping the family together. It was also her late husband's wish:

> I thought I should not leave money. That is why I had this house built ... Since I had the house built, I do not have anything left – some savings and monthly pensions that is all. The ownership of the house had already been transferred to my second son [who was living separately], so I do not need to leave a will. I do not want any nasty conflict among our children by leaving money. Leaving the house is safe, I thought. But it may not be secure even to leave a physical structure these days because some people even cut and take a piece of the Great Wall!
>
> (Mrs Matsuo, aged 77)

Sometimes, whether or not children owned a property dictated their decision-making process. Older parents were much freer to use their assets in old age if their children had already established their own home independently. Indeed, older parents fully performed their role in helping their children establish their own independent household through funding their higher education,

Analysis of perspectives on inheritance

helping with house purchases and so on. The strong sense of parental obligation perhaps attached more to investment in the process rather than at the end (with bequests). Among older British informants, however, financial performance, social position or family composition of children did not usually influence older parents' decisions regarding bequests. Irrespective of children's ownership of property, the fundamental idea was 'treating children equally'. If passing housing assets intact was the Japanese response to family solidarity and continuity, 'equal division of assets among children' was similarly important for British respondents to maintain their family relations and thus solidarity. Going back to Mrs Porter:

> You find this. When a will is settled, you find it at weddings, funerals, don't you? It brings all your emotions to the surface, when you are settling a will. Yes, things that have been bottled up for years or put up with, it all comes out when a will is being settled, and I do not want that between my two daughters, you see ... Well, I want to divide it equally between my two daughters ... Amy has this big mortgage so she has not much money behind her, perhaps I am thinking of that, you know. Whereas Susie has not got a mortgage at the moment you see [she was living with Mrs Porter at the time of interview]. I would not want to cause trouble between them. Knowing what happens [to my mother-in-law], going back to the will again, how it can split up families you know.
>
> (Mrs Porter, aged 71)

Helena, Mrs Davies' (aged 70) daughter, comments on her mother's attitudes towards such strict fairness in the context of family relations:

> I do not think she had that easy a relationship with her own brother. He died in an accident in 1985. And I know there was quite a bit of jealousy between them. So maybe my mum tried not to be like that ... My grandparents lived the other side of town. And they lived in a flat and my mum very much wanted them to move closer to us. You know for practical reasons as much as anything else. And there was a new sheltered accommodation being built and she wanted them to move and they would not. I think they would not because they did not want her brother to feel that they were moving closer to her and not closer to him. There was a bit of friction there. And maybe that is why she tries to be completely even with us and not favour one or the other. So I think that is probably why.
>
> (Helena, aged 39)

Such notion of 'fairness' in relation to the generational contract will also be analysed in the following chapter.

When unequal distribution of assets was mentioned by the British informants (and in some cases in Japan), it was often based on *family welfare*. Property transfer

Analysis of perspectives on inheritance

sometimes appeared to function to compensate a less fortunate child in the family, often without housing assets:

> This house will go to my youngest son, because the other three have their own house. I thought he and his family would move into this house, but I am not sure anymore because of his job [away from his home town].
>
> (Mrs Hirai, aged 72)

> Everything goes to my daughter. I have told my son about this, but he is okay over there [in the US]. He has done all right ... But I want Jennie [aged 53] to be safe. She has diabetes and you never know what is going to happen with that. If she ever had to give up work and that, you know, then it is a job to manage. She needs the money.
>
> (Mrs McKenzie, aged 82)

Bequests beyond children

Children are typical beneficiaries of family wealth in both societies, although not every child was necessarily mentioned by the older Japanese respondents. As discussed in Chapter 3, this is consistent with other research in Britain and Japan (for example, Finch *et al.* 1996; Rowlingson and McKay 2005). In Britain, when people outside the stem family were mentioned, it tended to be when they were left something sentimental or symbolic. On the other hand, non-stem family members were hardly ever mentioned as likely beneficiaries of assets by the older Japanese respondents.

Including grandchildren in a will can be a matter of debate in some families as the following dialogue between Mr and Mrs Clarke indicates. Mrs Clarke was preoccupied with the importance of fairness to acknowledge every member of their extended family as well as the value of timely support to younger generations with a realistic amount, while her husband was reluctant to alter the will to accommodate each life-course-related change. He expected their children (parents of the grandchildren) to look after their offspring, as they themselves had done:

> Mrs Clarke: '[We made a will] before Josh [a great-grandchild] was born. We have not got Josh on the will. I keep saying to you we should put our great-grandson on the will, shouldn't we, really ... '
>
> Interviewer: 'Who is on the will? Your daughters?'
>
> Mrs Clarke: 'It is just the daughters. It is just that it ought to be shared. If anything happens to Roy [husband], it will go to me. And vice versa ...'
>
> Mr Clarke: 'Well, all assets will be shared three ways with donations to the grandchildren'.

Analysis of perspectives on inheritance

Mrs Clarke: 'We mentioned the grandchildren, yes. But we have not put Josh, like the great-grandchildren, you see. So, I mean, I worry about that. I think, well, we should do that, you see ... I think I would like to up it for the grandchildren. You know, give them more because I feel that the daughters would be ... they would all be alright with what they got and that. So I would like to give the grandchildren more'.

Mr Clarke: 'But it means altering the will. It is up to their mothers to give them something ... This is how silly it is getting. We give something to the children and the grandchildren and we have got a great ... What about the great grand ... I mean, how much further does it go? If their own parents cannot give them something out of what they get, if they get anything ... '

(Mr and Mrs Clarke, aged 81 and 79 respectively)

This is an interesting theoretical question regarding how wealth trickles down over the generations. Mr and Mrs Clarke have three daughters – Maria, divorced with two children and a grandchild; Denise, married with children; Emily, divorced without children living in a rented flat. According to Mrs Clarke's view, Maria, who had more children/grandchildren, would be a winner, receiving more inheritance in total for her line of the family and despite her tenant status, Emily, who had no children, would receive the least. But according to Mr Clarke's view, all the assets should be distributed three ways and if Maria passed some on to her children and beyond her own share would be smaller. In this case, the daughters' housing tenures were not considered in the older parents' decision regarding inheritance.

Fairness and coping with life-course changes

Fluidity and the unpredictable nature of an individual's life-course influence people's decisions in bequests in contemporary societies. Unlike some other studies on intergenerational relations, in which multiple generations were interviewed simultaneously (see, for example, Wade and Smart 2004), this research was conducted in two phases over several years (interviews among the older informants in 2002–3 and their adult children in 2005). In some cases, transitions after certain 'events' such as death, re-partnering, a move and changes in health and financial status altered people's perceptions and plans regarding family relations and inheritance prospects. The research observed a variety of life-course events, and subsequent changes in both generations. Certain circumstances may significantly alter a decision, while inheritance could still be made according to customary tradition regardless of children's circumstances, as many Japanese people used to and some still do.

'Treating children equally', which was a firm favourite intention among the British respondents, at least in principle or on paper (in this case, in a will), was

Analysis of perspectives on inheritance

partly due to such changes and unpredictability in a life-course. In relation to will making, this echoes with Rowlingson's (2000) findings regarding the reason for people not making a will. Inevitably, some people may extensively rewrite their will each time a life-course change occurs, but like Mr Clarke above, the majority would rather not. If it is difficult to keep up with changes, for example, the view may be that it is best to keep the will simple, considering the unpredictability of how things may change in the future.

In this scenario, it is fair and easier for parents to give equal share to each child, and if necessary, they hope children would adjust the levels of wealth by making some transfers informally among themselves. A change in one child's economic fortune could make it possible to reverse the asset status between siblings by the time they retire, and the parents do not know what else may happen in the future.

> You would probably divide it equally in the will but hopefully you would somehow through conversations or show the other child let's say that if their brother or sister was really hard up they would be generous towards them, one would hope. I think you know maybe that is the only way one can do it. Because I think it is very difficult unless right from the start you know you could see there is a difference. To sort of you know be favouring one rather than the other.
>
> (Penny, aged 54)

Some parents are conscious about an equal distribution of assets among their children, and try balancing out the expenditure and investment on their children over a life-course so that an equal overall transfer of assets can be achieved. Similar cases were found in both societies. In some cases, despite the equal opportunities provided by parents, some children did not take the advantage or did not see successful outcomes. Although funding higher education, for example (which is often provided in full by Japanese parents), does not always promise children better life chances in the future, it is a substantial investment on the parents' part and is worth recognizing. Many respondents were very conscious about the equal distribution of investments in their children's upbringing:

> Toshiko is a full-time nurse, and has two sons (both married with children living in the same city) and an unmarried daughter working in Tokyo. Her daughter spent five years to obtain a music degree and the parents paid not only for her regular tuition fees and living expenses but also for additional lessons, related travel and such. Toshiko spent a 'small fortune' on the daughter, and felt she had done enough for her already. The eldest son took six years to graduate from his university, which the parents fully supported. While he and his wife were lodging in their house for two years,

Toshiko saved up all the rents the young couple contributed and gave the money back to them when they bought a second-hand house. Now, her concern was towards her youngest son, who had been working since he finished high school. He met his wife, married early and had children. Toshiko was feeling almost guilty that she had not helped him enough. She expressed her consideration of eventually leaving him their residential property.

(Toshiko, aged 55)

Jackie has two sons, both in their early twenties. One went to university and the other did not. 'Peter went to university and we paid for him three years in university so he came out owing no debt, no nothing … David, somebody wrote his car off so we bought him another car. So you know, you try to keep it as fair as you can'. Jackie paid David's insurance, and paid for all the furniture, all the carpets, the cooker and everything for his new flat with his girlfriend. 'You try and balance it out so … I mean Peter's probably, on all still probably have had more than what David has. But you know I am sure there will be something that David will need later on'. She was planning to split their assets 50:50 to her two sons.

(Jackie, aged 46)

As the above cases indicate, parents are similarly sensitive to the needs of their children across their life-course, and provide help on an as-needed basis. However, Toshiko's willingness to pass her housing assets intact to one child is firmly embedded in specific cultural practice.

Increasing rates of divorce, separations and remarriage in the past 30 years in these societies mean an increasing number of people have become part of what are called 'complex families' (Silva and Smart 1999; Williams 2004). Although marriage is still a popular institution in Japan, divorce rates have risen from 0.7 per 1,000 of the population in 1960 to 2.25 by 2003, according to vital statistics of the Ministry of Health, Labour and Welfare. The UK saw a large increase after the Divorce Law Reform Act 1969 but currently the divorce rates are in slight decline, at 12.2 divorcing people per 1,000 of the married population in 2006 in England and Wales, according to the Office for National Statistics. To a certain extent, the housing effects of divorce can be offset by remarriage, new co-habitation or a return to one's parental home. In Britain, however, it is evident that many individuals leave owner-occupation each year due to divorce (about 60,000 in the early 1990s), which results in the expansion of smaller households, often in other tenures (DETR 2000). The effect of divorce and the formation of step-families on wealth accumulation and inheritance through re-partnering is thus potentially significant in contemporary society.

Analysis of perspectives on inheritance

In terms of gift giving and inheritance, stepfamily dynamics appear to be different in different circumstances, depending on whether one has natural children and grandchildren or whether it has come about as a result of death or separation (Dimmock *et al.* 2004). It is therefore difficult to generalize the strength and quality of such relationships, and thus to assume likely practices. Even in Britain, evidence is so far inconclusive – it is still too early to assess the real impact of current demographic trends on inheritance practice. Finch and Mason (2000) found that the attitudes of respondents who had divorced and remarried revealed a desire to ensure that the assets did not pass down to someone who had no blood relationship to the donor. However, there exists a dilemma between protecting the blood line and the principle of fairness. With increasing divorce and re-partnering taking place in people's mid and later life, it is also possible that matters of partnership (especially forming later in their life), with respect to emotions and issue of property, are increasingly separated (Dimmock *et al.* 2004). In Japan, with the new pension reform in April 2007, divorce after retirement is expected to increase. The control and destinations of family wealth are thus increasingly uncertain in the ever-changing family structure and formation.

There were some interesting observations found in both societies, especially among the younger generation. In Britain, marital status of children does not usually affect the bequest plans of older parents, although some have provided extensive support in all areas of financial, emotional and practical support and accommodation at the time of crisis. Some of the younger generation told more interesting stories, especially those who were in complex families themselves or by anticipating such a scenario. Ben has a grown-up son from his previous relationship. He brought him up as a single father. By combining their individual wealth through the sale of their properties, Ben and his new partner were looking to buy a family home together in London. They have a baby in the new relationship:

> We have not discussed it [how to leave their assets] in detail. But I do not think she would be ... my ideal would be can you just kind of assume him [his son from the previous relationship] as yours as well, and we will just leave equal amounts to both. Because I do not think ... and she could rightly say 'well, he's got another mother', although his mother does not really have any money. She lives in a council flat and does not have any savings of any sort. But she might rightly say 'well, he is not my son, he does have other family on the other side'. And um ... we have to discuss it. But it is more complicated. So I may ... if she says 'well, I want to leave most of my money to our daughter'. I may compensate by saying 'well, I want to give a certain amount more then to my son to compensate for that'. It gets complicated but ...
>
> (Ben, aged 46)

Analysis of perspectives on inheritance

Some female respondents voiced their concern about protecting their assets from leaking out to a third party in case of re-partnering after their death. Wills are individual legal documents, although there was a great degree of consensus among couples in their wills. However, living in a more dynamic society, some younger informants made more pragmatic decisions and built in a different statement in their will from their partner, anticipating some possible scenarios, which were largely for controlling the destinations of their own assets by protecting them for their own children. Sally's view illustrates this point:

> There is another bone of contention. I have made a will. We have both made a will, but my will is a bit different from my husband's will. Because I have lost two quite close people within the last five years. And their husbands have moved on and married again within 12 months. And I am adamant that no floozy is getting her hands on my money or moving into my house. So I have stipulated in my will as best I can that it is to go to my husband and the children and nobody else is getting their hands on it. Of course, I cannot stipulate that fully. But like when mum dies and we do come into money I will change my will. Now my husband is a bit angry with me, but I have discussed it with him before, because I was really really hurt. Because one of the persons was a colleague that worked with me here and a very dear friend, and all her life she had scrimped and scraped to put her children through private education, things like 'Oh no, I cannot afford a pair of shoes at the minute'. And he is spending money and having holiday and moved house, and within 12 months of her dying. I just thought 'I am not having it happen'. I have worked for my children to benefit, not for a third party to benefit. But he is not happy about that because he said if I want to sell the house and move somewhere else I am not going to be able to do that now because I have got to give the children ... I do not care how unpopular that makes me.
>
> (Sally, aged 48)

This point brings us to another significant difference between the two societies regarding the sense of rights and control over 'share' of household/family assets.

Will: Autonomy versus lack of control

> In every society there are men who control more of the food, clothing and other forms of wealth produced in the society than do other men. The word control is used deliberately: from many points of view the control of wealth is more important than its mere possession.
>
> (Homans 1942: 339, quoted in Hann 1998)

Analysis of perspectives on inheritance

As the above quote suggests, if the control of property means the control of family relations, owning property certainly provides an advantage to older parents, for example, to negotiate family support from the younger generation. We may assume that they act strategically to maximize their benefits. The research found significant cross-national differences in this area between Britain and Japan (Izuhara 2004).

The older British informants were more autonomous in their plans to dispose of their housing assets at death. In addition, there was a high degree of consensus among the informants about how and to whom to bequeath. Almost all elderly owner-occupied informants had a clear idea of how to dispose of their assets by means of a will, although its contents were not necessarily very imaginative. Wills play an important role under the British system, since the dominant principle is testamentary freedom: in its pure form, individuals are free to leave their assets without consideration of rights or obligations towards their families (see Chapter 3). Once people have realized their housing assets by moving into sheltered housing or care homes, and started reducing their accumulated wealth, talk of inheritance tends to become something more focused on individual belongings: for example, who receives which piece of jewellery, antique china or furniture.

Otherwise, among the home-owning informants, the common arrangement was to leave everything to the surviving spouse, and on their death for the assets to be divided equally between the children. In contrast, only one of the Japanese informants had a will. The Japanese civil law-based inheritance codes are based on the principle that certain family members, such as spouses and children, have the right to an equitable share of the family assets. However, since the traditional family system used to determine the ways in which family wealth was passed on, Japanese people are not accustomed to making wills. The question of 'how to dispose of your assets at death' was therefore often met by laughter, even among home owners:

> We do not own anything which requires a will. We have nothing. I have to trust my own children. That is the principle. If children betrayed me I would be very angry. That is why I do not think we need a will.
>
> (Mr Soga, aged 83)

> I only have this house [with land]. There is not much to fight over anyway. I do not need to worry about what happens after my death ... Law defines equal division of assets among children, so they can follow that. Those who are left behind should decide how they want to do it. People make a will to avoid a conflict, but there is not much to fight over.
>
> (Mr Nishi, aged 73)

Analysis of perspectives on inheritance

Such contrasting responses can be explained as differences in existing social norms and cultural practices rather than differing legal arrangements or level of assets which individuals own. For example, despite the stagnated housing market, older Japanese people still own a significant amount of savings (see Figure 6.2). The legal arrangements of property left intestate do not differ significantly between the societies: the principle is that certain family members, particularly spouses and children, have the right to an equal share of the family assets. Instead, it can be argued that the concept of 'rights' or the notion of 'contractual society' is less developed in Japan in many areas, and thus unspoken rules and informal arrangements in the family often override the legal prescription. In the past, the family system clearly defined the ways in which family properties were passed on and there was therefore no need to make a will. Many informants mentioned that a 'will was only for very wealthy people'. It is not uncommon that many adult children, especially daughters, 'voluntarily' give up, if often reluctantly, their legal share in favour of a 'successor' male sibling: in the late 1980s more than 40 per cent of the respondents who gave up their share mentioned this reason (Noguchi *et al.* 1988). However, writing a will may help guide increasingly contested expectations of different family members caught between traditional practice and the legal definition.

Many Japanese responses to their plans of disposal for their accumulated wealth were ambivalent, however. Such views were often a reflection of the gap between traditional ideology and contemporary family practice, highlighted by rapid social change. Throughout the interview, for example, Mr Nishi came up

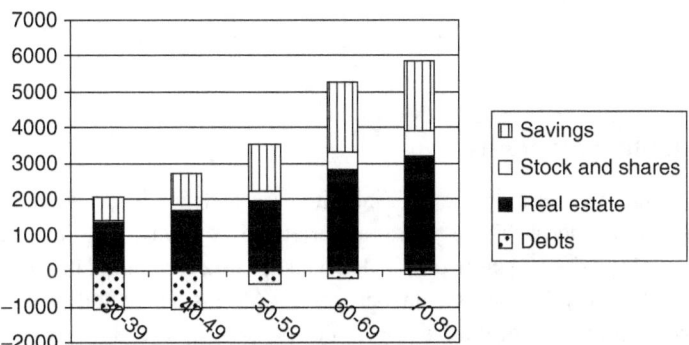

6.2 Value of asset per household by age group of household in Japan (¥000). Source: Arakawa (2003b).
Note: Survey conducted by the Daiichi Life Insurance Research Centre in 2002 among 918 samples aged between 30 and 80.

Analysis of perspectives on inheritance

with three different bequest possibilities. These scenarios contradicted one another but somehow captured well his highly ambivalent feeling, and also the gap between his expectations and reality:

> On one occasion, I was very angry to hear my eldest son planning to buy a condominium. I told him off saying, 'When I die, you got to do something about this house'. [He implied his eldest son should take over his residential property.]
>
> If I know my longevity, if I know I have five years and eight months to live, I would spend all my wealth before I die. I would re-mortgage the house and land, borrow money against them and spend them all. That would be great. Children are not pleased to inherit this house. If they were, they would come back and live here now. [This also explains an accidental nature of inheritance.]
>
> Law defines equal division of assets among children, so they can follow that ... [By this point of the interview, he realized he could not control his children's behaviour nor his longevity; he could not control the process. Then he settled the matter with a legal definition at the end.]
>
> (Mr Nishi, aged 73)

Turning to the views of the younger generations, what did they think about their parents' attitudes towards bequest? Many adult children wanted their parents to use their assets to lead comfortable lives for themselves. For unspecified cases in Japan, some children wanted their parents to express some sort of view over their assets and use it as a guideline when they settled their family wealth. This is a complex issue in relation to care provision so will be analysed extensively in Chapter 7.

How inheritance benefits

How the younger generation use their inherited assets depends partly on the level of assets they inherit. There were a variety of answers to this question cross-nationally. While a single physical asset, such as a residential property, may restrict beneficiaries for the use, financial assets (cash savings and investments) may provide more opportunities for multiple uses. For example, when the inherited amount was £200 from an aunt, Katherine bought a camera and called it 'Auntie May's camera'. When it was £2,000 from a grandparent, Louis had his windows replaced. In Britain, improvement in housing was one of the favourite uses for often 'unexpected' income. Putting inherited money back into housing, such as improvements in a property or paying off a mortgage, were often

Analysis of perspectives on inheritance

considered to be a wise and respectful option that eventually contributed to further accumulation of family wealth. When more substantial amounts were inherited, some informants used a range of ways, including savings, spending and *inter vivo* transfer. For owner-occupied adult children, some wealth skipped a generation, and 'unexpected' money appeared to be treated differently from regular incomes:

> I did get some money from my godmother who died last year. And in fact what happened was she gave all the money to my mother, who of course divided it among herself and her three children straight away. There was no reason for her to do it because the godmother had not made any provision for us at all ... when I got the money from my mum I gave my two children each enough to open an ISA [Individual Savings Account] straight away. Because I mean again it was money that I was not expecting to get necessarily ... we each were given something like £15,000 which was you know a good amount of money. So I gave three [£3,000] each to my children. And then the rest of the money saved and that has gone towards building a new greenhouse. You know so it is being used in the house.
>
> (Penny, aged 54)

In Britain, according to Rowlingson and McKay's survey (2005), answers to a multiple choice were evenly divided into those who had saved or invested money (41 per cent), and those who had spent it (37 per cent). People often kept smaller, more personal objects (27 per cent); 6 per cent paid off their debts, and another 6 per cent gave it away. In Japan, on the other hand, inheritance studies that are available are often preoccupied with the transfer of single physical assets (residential property), and the use here means whether or not beneficiaries would live in the inherited property. This suggests that in Japan inheritance is largely concerned with housing transfer. Koike's (2003) study in the Metropolitan region, for example, revealed that 68 per cent of inherited properties had been occupied by the beneficiaries for the past ten years while the proportion was likely to decline in the future (only 38 per cent of those who were likely to inherit in the future were planning to do so). Increasingly younger people mentioned 'paying for inheritance tax or a need to divide assets among siblings' (52 per cent); 'distance from the inherited property (implying the geographic proximity between the generations)' (32 per cent); and 'properties are too old' (32 per cent) as their reasons for not residing in their inherited housing. In Britain, in contrast, 75 per cent of properties inherited prior to 1995 had been sold and only a small minority were now living in what might have been regarded as the family home (Rowlingson and McKay 2005). This highlights the change in inheritance practices and its use in the context of social change, especially in Japan.

Analysis of perspectives on inheritance

Timing matters, as Pam and Ben explained how they were benefiting from inheritance given by their mother's early death. Their opportunities in terms of housing and employment choice had been opened up significantly. For Pam, her inheritance of £100,000 from her mother coincided with her husband's early retirement (with redundancy money of £80,000 plus £10,000 annual pension):

> We will pay the mortgage off, which is not big anyway. We will pay the car loan off ... [My husband] sort of suddenly thought I can actually do – the world is opened up so many opportunities – you know, I do not have to do a pressured job, I do not have to travel [for work] ... It has eased our whole feeling of stress levels ... we are so lucky, you know we are in our forties and we are very aware that circumstances have just happened and we are in a position that most people are not in until they are in their sixties.
>
> (Pam, aged 43)

Unlike receiving financial assets, some younger informants found inheriting residential property could be a burden. Such feeling is economic but also very much culturally bound. The research found such cases in both societies, but in Japan the situation could be more demanding because of the shorter life span of houses and the traditional notion of family obligations attached to property transfer. Ellie's husband as the eldest son is likely to inherit his parent's house:

> [The house] is absolutely beautiful but it is in the middle of nowhere, the nearest town is about eight miles away. So it is not practical when you are looking for retirement to sort of make things easy ... but they [his parents] would love it to stay in the family because it is part of them ... if my husband would be left the house, he would then be expected to I think pay half to his brother, it would not be just that he would have the house. I am sure again with them, the assets would be equally split. So if my husband had the house we would have to sort out that he would have a financial means.
>
> (Ellie, aged 46)

In this case, the location of housing made it problematic. They could not liquidize the house (because 'the parents expect the house to stay in the family') nor live in it as a main residence (due to the inaccessibility and little local connection), and would have to pay a half share to his brother, which would also be a significant financial burden. It is difficult for two such contrasting cultural expectations (residential property transfer and equal division of assets) to co-exist.

In Japan, some informants anticipated such a daunting scenario of having to rebuild an old parental home in the future. For example, Yuki (aged 41) felt much pressure from her parents-in-law (in exchange for their childcare support) to rebuild their house, and eventually to look after them in co-residency.

Analysis of perspectives on inheritance

Moreover, the research highlighted an issue regarding single, co-resident daughters in terms of their financial security and housing destinations. This is an increasing concern in Japanese society, reflecting current demographic trends including a decline in marriage (27 per cent of those in the 30–34 age group were single in 2000) and an increase in older parents living with unmarried children (see Figure 6.1). Some older parents mentioned their intention of leaving a residential property to their daughters (rather than to their son or dividing the assets equally among children), because they felt responsible for their future. Mr Nishi's daughter, Keiko, had been divorced since I interviewed her father and she was now living with her parents. With her low salary it was impossible for her to establish an independent household after divorce, let alone buy her own property. Keiko's view on property transfer was rather more pragmatic:

> My two brothers now own their own condominiums. But it is impossible for me to take over this house. It is an old house and will require constant maintenance. It is OK to live in at the moment, but if I am in charge of the house, there is property tax, maintenance cost, bills and so on. There is more than just owning the house ... It is not practical for me alone to live here. It is better if my brother moves in and I find somewhere else.
>
> (Keiko, aged 38)

By the time she was interviewed, both her brothers had purchased a condominium and she was not happy about that either. She thought it was a statement that the eldest brother was no longer planning (or never planned) to come back to live with his parents. She did not expect to receive a larger share of her father's assets; instead she would rather see her eldest brother fulfil his family responsibility. This was likely to add further uncertainty to Mr Nishi's already ambivalent views on property inheritance. Now having a divorced daughter at home, it has become uncertain what Mr Nishi would do with his housing assets.

Strategies for using housing assets in later life

Turning housing assets into an income stream has become a popular debate in many industrial societies to help older people have more financial security in later life. In Japan, this discourse was born largely out of the fact that a significant proportion of financial and material assets were held by older people nationally (see Figure 6.2). Such a neoliberal approach encouraging self-help rather than dependency on state welfare is indeed evident in the British context. This is debated in the context of a current recession, labour market restructuring and demographic change, which combine to produce uncertainty in future pension funds and return from investments. To secure and supplement their income and to meet the changing needs of housing in old age, there are various

Analysis of perspectives on inheritance

options that older people could deploy to release part or all of their equity tied to housing (see Chapter 3). This section explores the views and experiences of those informants who made a choice to liquidize their housing assets in old age.

Under the current English system, the strategy older people often adopt, albeit reluctantly, is to use their housing assets to purchase necessary care in a residential or nursing home. This was the most commonly stated, institutionally manipulated choice among informants. Many home owners, however, were likely to defer their decision and would try to remain in their own home for as long as possible, using a combination of informal support and private services. In Japan, in contrast, a move into supported housing was only an option for a minority, reflecting the underdeveloped care housing sector and the current condition of housing in the private market. Such private purpose-built housing for older people tends to be occupied by those without children (and thus without family support and obligations), but the research also found a number of cases where those who had children had made a positive decision to move into such housing. In such cases, the decision to break the 'generational contract' by liquidizing housing assets came from either the constraint of cultural practice or the liberation from it. This highlighted a strong gender dimension in family support structures. In any case, 'children being well established' was often a condition for them taking this option:

> My daughter is married to an only son whose mother was also widowed. It is right if his mother lives with them because it is her son. But it is not logical if I who married off my daughter expect to live with them ... I knew I was going to be on my own when my husband died and that is the reason why I chose to come here.
>
> (Mrs Adachi, aged 79)

> Men have a retirement age and can pursue their dream when they are free from a job. But women cannot retire from domestic work. I wanted to be liberated from the kitchen when I became 70 ... It took me two years to persuade my husband to move into this supported housing [which has a dining room for group meals in addition to an individual kitchen in their flat] ... Neither of my children had any intention of coming back so it did not matter whether we kept the house or not.
>
> (Mrs Yamaguchi, aged 81)

In both societies, many plans mentioned by older informants in the initial fieldwork were altered during the course of the research, and a housing-related move was significant in this context. This indeed explains the fluidity and unpredictable nature of an individual's life-course in contemporary societies. In some cases, transitions after certain 'events' altered people's perspectives on family

Analysis of perspectives on inheritance

support as well as plans for bequests. Many such life-course events are influential factors in transforming intergenerational relations, sometimes adding strains on the family. In one case in Japan, worsening health condition such as discovering the re-emergence of cancer prevented one informant from moving closer to her only son. In Britain, Mrs Davies (aged 70) was one of the informants willing to remain in her own house (four-bedroom terraced house) as long as possible, but had moved by the time her daughter was interviewed three years later.

An inter-tenure movement from owner-occupation to tenancy could be an attractive option for some who view owning a house in old age as a liability in terms of maintenance and property tax. In Japan, for example, given the fact that 80 per cent of older households are owner-occupiers and that the average size (floor space) of an owner-occupied house is almost three times larger than rented accommodation (Management and Coordination Agency 1998), the potential impact of liquidizing housing assets of older people on the rental market would be significant. By trading down, the current likely mismatch for empty nesters of household size and their living space, considering different stages of the life-course, could be rectified. The Ministry of Land, Infrastructure and Transport views such redistribution of housing stock as a vital and effective way of meeting the needs of households in different stages in their life-course, and is promoting new initiatives such as the establishment of support centres for movers and a sub-lease insurance system (*Gekkan Fudosan Ryutsu*, July 2003: 84–6).

Trading down could be the only financially viable means for older people who were willing to move from a stagnated neighbourhood to a popular, up-and-coming urban area. There were a few informants, both young and old, in the British fieldwork who mentioned the difficulty of older parents moving closer to their adult children in London, the Cotswolds or elsewhere due to the hierarchy of housing markets, and within the city due to property hot spots and cold spots. According to Helena (aged 39), there were multiple factors and opportunities that influenced her mother's move within the city, including vandalism behind her backyard, and her scheduled hip replacement that would make it difficult for her to cope with a two-storey house. Mrs Davies decided to move to sheltered housing, a 15-minute walk from her daughter. This intra-sector move within owner-occupation downsized her accommodation to a manageable level (a two-bedroom flat on the ground floor), but moving to a more desirable location did not release any of her housing equity; instead she needed to top up the money from the sale of her four-bedroom terraced house. This move appeared to have enhanced the quality of her life, and would enable her to remain independent for longer, while keeping the housing equity intact (for inheritance). However, Mrs Davies' move would now constrain the younger family's housing destinations. Helena was contemplating the choice of schools for her children (aged three and eight at the time of the interview) against house prices in such areas. Many of the parents she met in the playgrounds were moving to where there were some good secondary schools. Helena had a few more years before serious decisions needed to be made, but said: 'I do not see that I could now move anyway because mum moved to be closer to us ultimately'.

Similar cases were found in Japan regarding a move to live closer to children. The motivations behind the move were often altruistic; however older informants had no intention of becoming dependent on their children, but their move reflected more a willingness to help their adult children with childcare, to reduce the children's burden of travelling a long distance to visit them, or simply to see them more often (greater emotional support and companionship). When an inter-tenure move was out of reach owing to the hierarchy of the housing market, one Japanese informant was planning to move to public rental housing and expressed his concern over the move eating away his accumulated assets:

> Mr Watanabe's only son lived in Greater Tokyo: 'We are planning to rent public housing. It is not possible to buy another house with the money raised from the sale of this house. Houses cost two to three times more there ... We will sell this house and take the money with us. That will supplement our pensions but hopefully we would rather not touch the money ... We may need money for healthcare, so we need to keep our savings. It is a worry.
>
> (Mr Watanabe, aged 70)

Conclusion

How do older parents plan to pass on their housing wealth to future generations? The responses to this question illuminated significant cross-cultural differences in the inheritance practice between the two societies. Strong consensus around fairness and treating children equally were voiced by the British respondents, while mixed views favouring contesting one-child succession and equal division highlighted the process of social transition in Japan. Many Japanese respondents were still planning to leave their property to one particular child, intact, even if they lived in separate nuclear households. In contemporary society, such traditional practice was sometimes considered by their adult children as unpractical or even burdensome, especially when there was a conflict in lifestyles and choice of residence, when children did not have sufficient financial means to maintain the property or when reciprocal support obligations were attached to the property transfer.

Another significant difference was around the control over the accumulated assets by the older generation. While the British respondents exhibited strong autonomy in controlling the disposal of their assets by making a will, their Japanese counterparts were less preoccupied with its destination after their death. Many commented that their children would decide what to do. However, where the norm of practice has been diverse and is still shifting, such informal arrangement could become a source of conflict among family members due to different expectations, roles and contributions within the family. In terms of the impact of inheritance on adult children, the timing is crucial, as some informants highlighted in the research. The amount of inheritance also dictates how inherited wealth was used and a combination of methods (save, spend and pass on to their children) was mentioned. Inheritance did indeed find a route back to housing.

Finally, changes and uncertainty in an individual's life-course influenced people's decisions and the decision-making process in bequests. Changes in family formation, economy and housing markets, and individual health and financial status, all make it difficult for people to plan precisely how to dispose of their accumulated assets. Many strategies were deployed to cope with such changing circumstances but the underlining philosophy often appeared to be not to compromise the benefit of the family.

7 Rethinking the 'generational contract' between care and inheritance

Introduction

This chapter examines preconceived ideas regarding the 'generational contract' – how family members are bound to exchange various types and levels of support across generations over their life-course (see Chapter 2). In particular the chapter explores whether there is a particular link between the provision of care and inheritance in the micro-level interpersonal relationship within the family, using the voices of older parents and their adult children in Britain and Japan.

The rationale was largely derived from the cross-cultural differences in inheritance practice in the East and the West (see Chapter 3). For example, in Japan there has traditionally been a clear pattern of exchange between older parents and their successor sons, and conventional co-residency facilitated such processes of intergenerational transfer. This was previously a legally bound contract until the introduction of the new civil code immediately after the Second World War significantly democratized Japanese families. However, such traditional arrangements can still be observed in many families, especially those in rural communities. And accelerated social and demographic change for the past three decades has started to raise questions regarding the feasibility or even desirability of continuing such practice. In Britain, where independent living has been the norm among older people, family reciprocity and obligations have been organized differently – previous research has already confirmed that the link between care and inheritance is obscure or negligible. According to focus group interviews, 'generally people did not approve of linking care with inheritance. Most people were uncomfortable with the idea of linking any financial reward with the provision of care' (Rowlingson 2004: 38). Today, the issue regarding the delivery and finance of care for older people goes beyond individual and family obligations and cultural practices, and has become central to policy and politics in both societies. This issue is being debated in relation to personal asset accumulation.

In this chapter, a few themes that emerged from the interview data with the two generations will be explored, particularly in relation to the generational contract exchanging care and inheritance: independence and family support; inheritance as reward or recognition; fairness and equality among children; investing in the process or outcome; and subjective or objective views regarding the practice. The younger generation's views are drawn on substantially. Furthermore, in Japan, the year 2000 was a watershed, when the government introduced the fifth social insurance scheme on long-term care following health, pensions, workers' compensation and unemployment insurance. This was largely a response to increasing care needs in a rapidly ageing society and has significantly altered people's views and attitudes on long-term care. Based on interviews with older people and service providers in 2002/3, the chapter assesses the early impact of this new scheme on the generational contract. Finally, it concludes with an illustration of the complexity and dynamics of family relations regarding the care–inheritance nexus, based on a case of one particular Japanese family.

Independent living and the 'generational contract'

We begin by considering the perspectives of the older generation regarding family support and long-term care. In the Western literature, care is now increasingly considered as a 'relationship-based activity' that needs to balance the rights of both people involved in the caring relationship (Lloyd 2000; Barnes 2006; Fine 2007). Finch and Mason (1993: 58) highlight this complexity in caring relationships: 'issues of power and control are closely intertwined with the negotiation of the balance between dependence and interdependence'. Whether a form of dependence or a process of collaborative assistance, many commentators believe that without reciprocity, a caring relationship can become a one-way activity, an oppressive practice which may lead to excluding the perspectives and neglecting the rights of the recipient (see Lloyd 2003; Fine 2007).

'Living independently without becoming a burden to their children' was a wish commonly expressed by many older British informants. Despite the tradition of strong family support, a significant number of older Japanese informants also shared such a view. The demographic location of family wealth, and thus the likely flow of support between the generations in contemporary Britain and Japan, shape their attitudes towards the receipt of support and bequest plans. Among the owner-occupier informants in both societies, older parents tended to be the net providers of family support. Even for older people who were fiercely independent, however, prolonged longevity could require a certain level of support and care in the later stage of their life-course. In this scenario, older parents could adopt certain strategies for receiving long-term care from various sources. The question lies as to whether they used their assets including those invested in housing to negotiate family support, or whether they would rather use them to purchase services available in the market or elsewhere. The newly introduced social insurance scheme on long-term care in Japan has created more

Rethinking the 'generational contract'

opportunities for older people to exercise their options beyond conventional informal family support.

In Japan, the value shift has been accompanied by the increased economic independence of older people that is also evident in the decline in co-residency. Not surprisingly, the generational contract (of reciprocal family support) still appeared to be alive and well in the case of co-residency, but even co-residency has lost its traditional resonance, emphasizing eldest son succession. At the same time, the expectation of family support, especially with regard to daughters-in-law providing the bulk of the heavy end of nursing care, is increasingly becoming diminished or seen as undesirable. Policy shifts towards the wider socialization of care have confirmed and accelerated the breakdown of this preconceived contract (Izuhara 2003a, 2004). Today, the new scheme has made it possible for older people to achieve a higher degree of independent living even within extended family living arrangements. In this context, it is fair to say that the relationship between care provision and asset transfer within the family is becoming increasingly divorced in contemporary Japan.

In Britain, older parents' desire for living independently was well highlighted in this research. Even in the case of co-residency, for example, an attitude illustrated by Mrs Gaskin (aged 78), who was living with her daughter's family, was that 'if she reached the state where she needed extensive nursing care, she *must* leave the house'. In co-residency, she retained a certain level of privacy and independence (see also Chapter 7), and she had no plans to use her accumulated assets (invested in the shared residence) to further negotiate family support from her co-resident daughter. The older mother's *feeling* of independence was supported, however, by her co-resident daughter, Mary (aged 51), and her family's sensitive and discrete approach to care:

> When we go to bed we check that she is okay obviously and come in during the day. So there are lots of points during the day when she is being checked on if you like without it being obvious that is what we are doing. I can do things like – her sight is not terribly good any more so she does not actually know when she has cleaned things very well. So we go in and clean her bathroom when she goes out, so that she does not realize that we are doing it. I think her pride would be quite upset if she thought we needed to do it.
>
> (Mary, aged 51)

This is a good illustration of the discrete moral and practical support put in place by the younger generations in society where the notion of independence is highly valued. However, Mary's experience has shaped her attitudes towards the generational contract with her own children:

> I actually find it [co-residency], although I would not have it any other way, I actually find it quite intrusive. And I think we have been in this house nine

Rethinking the 'generational contract'

years and I think I can count on the fingers of one hand the number of times I have been in this house by myself. So I do not ever get solitude. And if I come home and want to have a day by myself I cannot do it. And if I come home from work having had a bad day I have still got to go in and be polite and nice. I find it quite difficult. Just sometimes I just do not want to have to talk or do not want to have to answer a question. And I do not think I would put my own children through that. Having experienced it myself.

(Mary, aged 51)

Overall, the cross-national research found no strong link between bequest motives and family support for long-term care among the older informants, which was initially considered as a vital exchange tradition, at least in Japan. The 'willingness to leave more assets for children who looked after them' – a commonly quoted survey result in recent years (for example, *Kyoto Shinbun* [newspaper], 28 March 1993), which originally inspired this research – is thus somehow misleading. There were instead multifaceted responses to the statement, 'people who look after their parents should receive a larger share in inheritance than those who do not', in the research. Among the older informants, housing assets were generally not used to negotiate family support, but common and shared responses in both societies were that people did not want to become a burden on their children in old age, but to a certain extent were willing to leave something to them.

In Japan, apart from the cases of co-residency which were built in implicit expectations of family support reciprocity involving asset transfer, if one child helped more or provided support on a more regular basis without sharing accommodation, some older informants tried to compensate for their contribution on the spot rather than let them accumulate any advantage and leave it to them in their will. A similar approach was found in both societies:

> We try to be fair to our children [although they are willing to leave the house to one particular son]. Our two daughters help us every Sunday, but we will give something each time. Just a token of appreciation. If they bought something costing ¥2,000 we give ¥5,000 instead, for example. They drive to come and see us so we just want to show them our gratitude.
>
> (Mrs Nanri, aged 82)

> [My daughter] drives down once a week for a day. She spends a day with me … and we either go out to places or she helps me with some of my paperwork and tax returns and things like that … [My assets] will be divided equally between my son and my daughter … [Interviewer: Do you consider rewarding her for her help?] I do, I do. Financially or just payment each week. She had to outlay a certain amount to travel down here … running a car is quite a costly business. Yes, I do satisfactorily reward her for the cost … It would not

> be fair otherwise, would it? But it is not my way of payment for what she does for me, if you understand. I do not think she would have that anyway.
>
> (Mr Dawson, aged 85)

The unpredictability in their children's life events, as discussed in Chapter 6, was also part of the reason why people preferred this approach. Among the British respondents, when treating their adult children, fairness and equality came across as more important conceptions than intergenerational reciprocity. In this scenario, fairness was often attached to material distribution, more precisely, the division of assets among children. According to Turner (1986), there are three different conceptions of equality. Equality of *opportunity* is typically defined as access to all social positions that should be governed by universalistic criteria. Such equality of opportunity, however, only makes sense if people start out from an equality of *condition*, including similar ability and assets. Equality of *outcome*, that all should enjoy the same standard of living and life chances, may be a more idealistic notion. The three conceptions are closely related to one another since, for example, the promotion of equality of opportunity among individuals, neglecting equality of condition, must result in an unequal outcome. Moreover, notions of fairness are often associated with ideas of equal treatment and the absence of bias, and there is a fine distinctive line to be drawn between fairness and justice (Sugden 1992). This confirms theories on fairness which tend to be associated with procedures rather than with outcomes.

Inheritance as reward?

The research among the younger generation revealed that inheritance was generally not seen as a reward for family support, but rather family support was viewed as a combination of duty, reciprocity and voluntary action. In the two societies, it was believed that people should not provide support to their family members for money, but the language that the respondents used to describe their motivations varied cross-nationally, highlighting the different 'cultures of care' that exist in the East and the West. As Chamberlayne and King (2000) argue, patterns of informal care emerge not only as the direct product of laws and public welfare systems, but as an ideological, cultural phenomenon in their own right. For them, such culture 'interweaves an analysis of action (both intended and actual), meanings (ideologically received and more personally derived), and patterns of social resources and relations' (ibid: 5).

> Sally, a full-time teacher, visits her elderly mother (aged 83) and has lunch with her every day; she has much closer contact with her mother compared with her five other siblings. The mother used to look after Sally's children when she went back to work full time. When her father died, the mother

suffered from depression and Sally moved in with her for three months with her children to look after her. 'It has never been a hardship for me. I have never done any of it because I felt it was my duty. I have done it because I wanted to do it'. She assumed that the assets would be divided equally among her siblings.

(Sally, aged 48)

Zena's parents moved to live closer to her after their retirement 10 years ago and thus she benefited from occasional childcare from them. 'It would be something that I would want to do for just the love and respect that they have always shown us'. She assumed that the assets would be divided equally between her and her brother.

(Zena, aged 42)

Mika is a full-time teacher. Although she lives with her mother-in-law, her own parents were the ones who provided extensive childcare for her children. Her parents used to drive half-way to pick up her children, take them to the nursery and looked after them after the nursery. '[Among the siblings], my weight towards care for our parents would be greater. I must do that – it is my responsibility'. She expected her parental assets to go to her brother who lived next to the parents because she has enough from her husband's family.

(Mika, aged 44)

Having herself experienced the intensive nature of support provision in co-residency, Mary (aged 51) viewed the caring responsibility slightly differently from other informants in the British context. For her, 'it is family. It is not a job but responsibility'. Again, she did not believe in any financial reward for her support, but her love and respect were underpinned by a strong sense of obligation – especially a close bond developed through her mother's prolonged widowhood since she was 10 years old. Instead, she appreciated more symbolic, non-monetary gestures from her mother to acknowledge her support. For example, her mother bought her a little tree for the garden when she came out of hospital.

There were also some traditional, gendered views expressed by the younger informants in both societies. Looking at the situations and profile of adult children in the family, it is often predictable who is likely to provide or is actually already providing certain types of support for their ageing parents and why. There may be different levels of expectations and actual provision between the East and West but the criteria used for speculating on what support is indeed similar (for example, geographical proximity, gender and existence of other competing commitments). It is often a popular discourse shaped by the social norm, but social policy also plays a part in shaping such practice. Some governments ration the provision of public support if certain members of the family are *available* and *capable* of providing support (for example, the proximity to daughter used to ration services in Britain; the availability of family support in Japan):

> Jackie is the only daughter of three children. Despite her younger brother living with their widowed father, she has a clear sense of obligations. 'No, I think [inheritance] should still be divided three ways because it is just expected that the daughter would do it [providing support] anyway, isn't it, really. Because you know, they are normally at home, aren't they, women'.
>
> (Jackie, aged 46)

> I do not expect my sons to look after me in exchange of our house. If so, they are no longer my children [blood relations]. It is just like an economic bargaining. I do not like such calculation. Regardless of financial reward, they should really help their parents.
>
> (Koji, aged 44)

Koji's view is traditional, defining family support as the norm in Japan. It is also pragmatic, however, when it comes to who should provide support to their own parents. Comparing his situation with his sister, who is married without children, is not employed and lives next door to him, he had no hesitation in expecting his sister to provide support to their parents without any particular financial reward:

> I am working and my sister is doing nothing. Even if she helped more, there is no reason that she received more inheritance. If I was not working, I could have done so. If I am not doing anything and she did it all, I would ask her to take everything.

This also captured well the stereotypical conceptions regarding care: care is free labour, and the status of married women is as non-wage earners but dependent on their family (husbands). The issue regarding the cost of care is further explored in the following section.

Another interesting finding is that those who provide more care appear to receive more in the process to reciprocate their inputs. In fact, there were many cases where the close proximity to parents also provided an advantage to children themselves if they were in good relationship. For example, an elderly Japanese mother used her daughter as a 'private chauffer' on a daily basis, but the mother was the one who bought the car for her daughter, paid for petrol and paid for the daughter's shopping as well as her own when they went out. The daughter may become a net provider of care in the future, but that would only reciprocate the gains that she experienced earlier in their caring relationship.

Furthermore, it is common for different tasks to be performed by different members of the family, which may complicate the process of 'fair' reward.

Paul (aged 63) and Sally (aged 48), for example, were two of six children who provided a higher level of support and supervision for their mother. There are particular explanations for their involvement, including their geographical proximity to the mother, Paul's retired status compared with his brothers who were still in employment as well as the compatibility of their personalities. The tasks that the siblings performed were organized around certain gender divisions of labour. While Sally provided daily input (visiting her mother every day) and more intimate support (moving in for three months when her mother was ill), Paul provided transport and managed his mother's finances. This case presents the dynamic nature of family support – different members have different caring capacity and commitment, and different types and degrees of support are provided. But such dynamics can also create disputes:

> Their youngest sister lives abroad and only visits occasionally. Her visits are financed fully by their mother. 'For her, mother is not 85, she is still 42. She arrives with the children and takes over the home. The whole house looks as though a bomb has hit it. She is only there for a week and then she goes away. And it takes us about three months to sort of get it all back and get mum back on an even feel again.
>
> (Paul, aged 63)

Reward, recognition and cost of care

Informal care can be an onerous task, physically and emotionally. In Britain, for example, only 'those who provide or intend to provide a *substantial amount of care* (20 or more hours a week) *on a regular basis* for another individual aged 18 or over' are eligible for public services (DH 2005). There have been increasing demands among carers for recognition for the care they provide as well as for improved material conditions. As Lloyd (2006) points out, however, there are problems in linking these two rather separate demands. If they are only given recognition without improved material conditions, informal carers will remain at risk of social exclusion, particularly in societies where workfare is replacing welfare in a policy trend, emphasizing the pre-eminence of paid work as a route into financial security and social inclusion (ibid.).

Provision of informal care is in many ways costly for family carers, although like other domestic work, it is difficult to monetize and thus to compensate for it financially. In Japan, the socialization of care under the new insurance scheme on long-term care has meant that the monetizing of such care work is now provided by 'formalized' paid care workers – the system for payment has been revised from having hourly rates for different tasks (for example personal support) to a more complex points system (certain points given for different tasks such as cooking, taking a client out for walk and so on). In this policy context, however, informal family carers continue to be unrewarded and thus not recognized for their contributions in monetary forms.

As well as working without wages, other costs are also incurred by informal carers, including real costs (for example, transportation), opportunity costs (giving up or reducing hours spent for paid work in the formal labour market), and also social costs (giving up time for other activities and possible isolation from peers and society). How do people weigh up the cost of care and their sense of love, reciprocity and responsibility to take up the task?

Some younger informants, in particular those who were actually involved in such caring activities personally or professionally, were much more realistic about the actual hardship and costs of care, and expressed more realistic views towards their responsibility.

As a practice nurse, Masami knows that caring can be hard work, and it is thus difficult to sustain the Japanese tradition in contemporary society. It is hard work for one person to bear all the physical side of caring single-handedly, as many women used to do in Japan. Among her siblings she is the best situated to provide support to her parents (that is, married without children, with professional experience in nursing), but she would still not do it for any reward. If she did, however, she would want some sort of 'recognition' from her siblings:

> If my brothers and sister want to express their gratitude towards me financially, I would be pleased to receive more money. But that sort of 'willingness or feeling' from my siblings matters [more than actual reward]... Caring is hard work but also rewarding in other ways.
> (Masami, aged 46)

> Many people say they do not want reward. They are doing it because they want to do it ... But the trouble is we are so aware, and it is no good pretending that money does not matter. Because we have realized that the fact we are getting this money [inheritance from her late mother] is going to give us freedom to do things that we could probably never have done. And freedom to perhaps give the kids ultimately something to help them. And I do think, I mean, if you have not got money, enough money, life is really hard.
> (Pam, aged 43)

While Pam focused on the value of financial reward (in her case an unexpected inheritance), Masami mentioned a different type of reward and thus reciprocal gains involving the provision of care. According to her, caring was hard work, but could also give emotional benefit and satisfaction by supporting the final stage of someone's life-course. She has witnessed such cases in the hospital.

There appears to be a significant social change of attitudes reported by the younger informants. Informal carers have saved the national economies billions of pounds (for example, £57 billion a year in the UK), but have generally been undervalued and thus felt 'exploited' in both societies. This goes hand in hand with the facts that carers are often women and on a low income, otherwise they would not take up the role as a carer in the first place. Their own earning is sacrificed, which may affect their financial health in the long run. There is an issue for those without a partner. More recent social and policy change has, however, meant that voiceless people – often women, on low incomes – are speaking up (see, for example, Carers UK 2005). Both (financial) reward and (socio-political) recognition are key conceptions for improvement, especially given the historical undermining of their contribution in this field.

In terms of financial reward in the name of inheritance, unless parents make their will specifically in favour of their family carers, their informal wishes and intentions are often difficult to materialize. This is in contrast with the opinion expressed by many older informants, that for any imbalance in children's circumstances, 'it is better to be sorted out informally among children'. Such vagueness in their decisions may be due to the fact that people often feel awkward talking about finances within the family. Family members could have different interests and opinions, and the informal intentions and wishes of older parents may not truly reflect the outcomes of their bequests if they relied on an informal arrangement. This issue is explored in the final section of this chapter using one Japanese scenario.

Under any circumstances, however, some people still feel uncomfortable treating children differently in their bequest plans. If there are some legitimate reasons why they feel a particular child needs to be rewarded or helped more financially, there appear to be a variety of ways of delivering it. Despite the private nature of financial talk, many British respondents emphasized the importance of open discussion for needs-based or reward-orientated transfer as a key to sustain 'family solidarity':

> In principle I think that would be OK [to give a larger share to a child who provided more support]. But I think it would need to be discussed openly with all the family, and everyone sort of agreed that this is fair. But I think that ... That would be okay because I know how demanding it can be ... [Interviewer: What about your children?] Again I would want to discuss it openly with both the children. And say 'look this is what I feel. I think it is only fair because of this'. And see what people feel. Yes. I do not think it is good to leave it as father of my brother-in-law did [In his will, he did not leave anything to his son. His son was wealthy and did not need money but he was very upset that he was cut out from the will]. It causes tensions and resentment'.
>
> (Ben, aged 46)

> Sort of thinking of my parents, they have been scrupulously down the line with us, you know. They have had two children and so we are treated equally. But if there is a different situation, like you know my sister's situation

> [when she divorced], then they will step in and help. But they will let the other one know. Which I think is fine. You know nothing is done 'Oh here is £50. Do not tell her' – none of this sort of thing. It is all very upfront. Certainly that is what I have grown up with.
>
> (Ellie, aged 46)

In Japan, where such a democratic notion of 'treating children equally' still needs to find firm roots in people's minds, a more discrete way of supporting a particular child financially is possible. For example, Michiko (aged 54) wanted to help her son who lived in the same city more than her daughter, but wanted to do it quietly, at appropriate occasions in his life-course.

Different scenarios, different responses

In both societies, different scenarios produced different responses to the question regarding the exchange of care and housing assets within the family. Some informants gave conditional answers to the statement: Those who provide more support should receive a larger share of inheritance'. For example, people mainly agreed with the statement in principle, but looking at it in their own family situation, it all depended on the situations of different members of the family. Moreover, some informants differentiated objective views from subjective circumstances, and others differentiated their own contribution from that of their in-laws.

The conditions that determined their views and thus influenced the legitimate division of assets were various, and again multiple factors influenced their views and attitudes. The financial status of older parents – whether they could pay for their everyday life and cost of care, in particular – matters, and that of their adult children, especially the financial circumstances of potential/actual carers, also counts. What level and length of contribution children have eventually provided may also need to be considered. If, as many commentators mentioned, family support is becoming more like *organizing* support provision using various available sources rather than *direct provision* of care by themselves, their contributions are viewed differently – and often it does not justify any extra reward for such an organizing role. Geographic proximity of children from their parents and other competing commitments such as having a young family or being engaged in paid work can also be considered. Another interesting criterion mentioned by the younger informants was a reciprocal element of support between the generations:

> If I provide more [support to their widowed father] living near my family [in the same city], I would not demand more since my brother who lives far away [in London] and has a baby cannot practically do so.
>
> (Pam, aged 43)

> Yes, I kind of agree with that [leave more to those who provide more support]. Perhaps things could be skewed in my sister's favour [who lives near their parents and thus potentially provides more support] but you know if it was, let's say, split 70:30 then I would feel a bit hard done by that. The reason for me not being there is because I am working elsewhere in the country ... She has received a lot of help in terms of free childcare, which I and my wife have not done, then it might make more sense just to split it straight down the middle. Even though she has perhaps been around to offer some support, she has also been earlier a net receiver of support.
>
> (Shaun, aged 39)

Mari was single and lived with her parents. She was not hard up herself but was sympathetic to those women without a well-paid job who were likely to be drawn into such caring responsibilities at home:

> If those who provided care are left without pensions, or if parents are poor and their son or daughter supported them financially, they should get more. It depends on the parents' living standard as well. If adult children sent money to parents monthly, whatever is left or saved should be theirs. You would not argue that you should divide that equally. But if parents are wealthy and a person like me lives in their house, I am not sure who is really supporting whom – so I would not demand more share.
>
> (Mari, aged 55)

> One of my husband's sisters is an occupational therapist and deals with the elderly. And there was the possibility that she might have moved in with the parents and done it [providing care]. Now she is in a situation where she has very little money. She is very poor. And had she decided to do it for, let's say, the two years that her father stayed at home, if he was to stay at home, then I think probably her father – because he would not think in those terms in any will or anything – he would not make any provisions. But I am fairly sure that the brothers and sister would have said, 'OK, Deborah has done this for two years, she must have more money than the rest of us' because of her financial situation. But if it happened to us and I let's say moved in and looked after my mother for say – and I would only do it in the case if it was months, rather than years. Because I still feel I have family responsibilities of my own. Because of my financial situation in relation to my brother and sister, which is fairly similar, I do not think there would be any question that I should have more money than they would.
>
> (Penny, aged 54)

Furthermore, given the different circumstances among siblings, some Japanese informants suggested there should be more emphasis on fairness in the contribution

Rethinking the 'generational contract'

process, rather than adjusting the levels of inheritance to achieve a fair outcome. Financial contribution by non-care-giving siblings could be made in the process to compensate carers' physical contributions:

> Yuki is a nurse and has two brothers living in Tokyo with their young family. Although their parents are highly independent, she is likely to help the parents if need be. 'I would compare the money they would send every month and my own labour contribution. Brothers might say "we would pay, so could you please go and help them!" Or they might say "we should all make financial support, but we would each pay ¥50,000 monthly and you need to give only ¥20,000 but can you help them?" I am pleased that both my brothers have a well-paid job and sufficient financial ability to do that. It would be impossible for me to provide both financial and physical support'.
> (Yuki, aged 41)

Subjective and objective scenarios

Interestingly, there was a clear pattern of responses depending on the informants being potential carers or not. Being objective, if the respondents were not likely to be involved in direct care provision, they tended to agree with the statement 'those who provide more support should receive a larger share of inheritance'. Their answer could be driven by their gratitude, generosity, guilt or even evasiveness. Indeed, inheritance could be a burden if inherited property assets held little financial value, were inconveniently located or were difficult to use or liquidize by beneficiaries (see Chapter 3), and thus some people showed no strong wish to receive it. Many expressed ambivalent feelings regarding the generational contract, and were aware that they would not be the ones who would provide care if necessary but also knew that voicing their willingness to give up family assets could place further pressure on potential carer siblings:

> I totally agree with the statement. I have no attachment to the parental assets. It is natural to assume that my sister will be the one who could provide support if necessary. It is fine to me in that case if she inherits everything. But if I say this out loud, it sounds as if I am placing a sole responsibility onto her.
> (Hideo, aged 38, Yuki's younger brother)

> Hajime is the eldest son but lives away from his parents while his younger brother lives next to them. 'The assets should go to the person who contributed the most. That will ease my burden. If I am given an equal share, I have to have some responsibility as well. But for me, it would be disadvantageous to take such a load given my job and lifestyle'.
> (Hajime, aged 45)

Rethinking the 'generational contract'

In Japan, despite the traditional practice of care provision by daughters-in-law, in the old system they were not usually rewarded personally. The decline in co-residency now meant a shift in expectations and modernized views among the younger generation. Some respondents in fact differentiated the reward element of care provision between themselves and their in-law counterparts:

> In principle, if my brother's wife helped, I would like to give her some sort of reward. Because she looked after *my* parents. But if I provided care, there would not be any connection [between care and inheritance].
>
> (Mika, aged 44)

On the other hand, if the respondents were potential carers themselves, they tended not to expect rewards for their contribution. There appears to be a complex set of explanations for this and feelings such as modesty (highly valued in Japanese culture), love, duty and responsibility, and reciprocal gains over time. The mixed feelings highlight cultures of care beyond the differences in welfare regimes.

Moreover, the research in Japan highlighted an issue regarding single, co-resident daughters in terms of their role as potential carers, their own financial security and housing destinations. For example, as discussed in Chapter 6, Keiko (aged 38, divorced without children) expressed ambivalent feelings over issues of care and residential property transfer. On the one hand, as a co-resident daughter recently returned to live in the parental home after divorce, she felt responsible for supporting her parents in the future, but on the other hand, she worried whether her financial status would allow her to do so. She needed to carry on her work to support herself, and leaving her job to look after them would jeopardize her long-term financial health as a single woman. This is an increasing concern in Japanese society, reflecting the current demographic trends such as decline in marriage. To support their unmarried child, some older parents in Japan mentioned their intention of leaving their residential property to the daughter (rather than dividing their assets equally among children). As also discussed in Chapter 6, however, it may be difficult for those unmarried daughters to maintain their inherited residential property. Sayuri, who is single and co-residing with her parents, also expressed a similar view:

> If I see it objectively, I would agree with the statement. Those who looked after their parents should receive preferential treatment than those who did nothing. But when putting it in my situation, I wonder if that is right. So far, it is difficult to say who has been looking after whom. I do not think I should receive more than my brother [who lives in Tokyo]. Umm, it is difficult to speculate. Now I am talking in principle, but if I were put in the situation and I were struggling to make ends meet – I do not know how I feel about

Rethinking the 'generational contract'

> it ... My father's pension is much better than my salary, so when he passes away and I have to look after my mother, it is impossible to maintain the same living standard.
>
> (Sayuri, aged 45),

Those co-resident daughters are usually in employment, and considering their own future it may not be realistic for them to provide an extensive level of care to their parents in the later stage of their life-course. Considering the current work pattern of women, it is not uncommon that elderly parents are left alone at home during the day, even in co-residency. In Japan, this fact partly provided a rationale for developing the new social insurance scheme on long-term care to fill the care gap vacated by the family.

Impact of policy shifts in long-term care

This section presents how policy shifts under the new social insurance scheme on long-term care in Japan have impacted on the generational contract (for more detail of the scheme, see Chapter 4). The qualitative research among the older informants and service providers was conducted in 2002–3 and evaluated the early impact of the scheme on family relations after its introduction in 2000.

Whether policy is a facilitator or a follower of current social trends, there has been a clear shift in people's attitudes and practices regarding informal care within the family. This is especially evident in contemporary Japan, where the transition has been more significant in recent years. With the introduction of the new social insurance, many older informants talked about the changing climate surrounding family care. The rationale behind the introduction of the scheme not only reflected a demographic and economic downturn, but the family was also on the verge of collapsing under the heavy burden of care, and the declining capacity of the family to care, partly due to the increased geographic distance between the generations and also increased female labour market participation (Peng 2002b). Many older people stated that they were 'the last generation having looked after their elderly parents [in-law] and also the first generation not being looked after by their adult children'. The shift sometimes happened without any choice. Some informants, however, also suggested that there were benefits from the social change – the shift of care provision taking place in Japan under the new social insurance largely influenced the conventional generational contract. Children were no longer the only or the most desirable choice of carers for current older parents:

> There are advantages of family care because we understand each other well. But it also depends on personality. Some people have great reservations

> [towards their own children]. In that case, it is easier to ask someone to do it as a 'job'. But I do not know how close I could feel to the person.
>
> (Mrs Yamaguchi, aged 81)

> Society has changed and everybody has a job now. If my children are willing to look after me, I will let them. But it is easy to ask for public home help since it will be easy to set a boundary. I would expect children to do things in a certain way and may feel frustrated if they do not respond to my needs.
>
> (Mrs Kita, aged 77)

Since its introduction in 2000, the new scheme has been successful in expanding the number of services and providers (for the early evaluation see Hiraoka 2002). The scheme has begun to include a much wider section of society as recipients, especially those who used to be excluded from public services due to their financial and family circumstances. Today, older people who have a sufficient income, including a high level of occupational pensions and assets, could also enjoy a better quality of life with formal services (not necessarily just with nursing care but also practical support such as cooking and cleaning) at lower financial cost.

The steady expansion of services over the first five years can be viewed as proof of the successful 'socialization of care'. For example, the number of home help uses increased from on average 3,550,000 times per month in 1999 (before the introduction of the scheme) to 5,390,000 times in November 2000 (immediately after the introduction) to 8,160,000 times in April 2004 (Ministry of Health, Labour and Welfare 2003). The most dramatic increase was seen in the numbers of group homes for people with dementia from only 103 homes in 1999 to 903 in 2001 and to 1,574 homes in 2002. The scheme has significantly benefited those people who had care needs and applied for services, while the life experiences of some others have remained unchanged. The following examples highlight contrasting cases of service users and non-users with fairly similar needs. In both cases, older couples required regular visits to clinics (although this particular service user had a worse heart condition):

> We will support each other until we die. We go out everywhere, shopping and hospitals together. Otherwise what happens if one falls? We hold hands when we walk up a hill ... I have not thought about using the care insurance. We do not want to rely on public welfare. We would rather help each other ... Because I do not drive, we always have to walk. When we went to hospital yesterday, we took a taxi one way and walked back. It took us an hour and a half. It was hard work – walked little and needed to rest, walked again little and rest ... I have heard about a 'care taxi' but do not know how to apply for it. A taxi one

way cost us ¥910 that is why we could not afford both ways ... When we step out of our front door, there is always a cost.

(Mr Soga, aged 83, non-service user)

The care taxi picks me up twice a week to go to a clinic. We thought the care insurance was only for those bed-ridden people. We had no clue because nobody seemed to know about the details. But when I got ill last time, my [co-resident] daughter was told by a mother of her student that we could apply for the services even though we were not bed-bound ... My husband used to drive me to hospital, so his burden has been lifted. My daughter was concerned about him as much as about me, and I also wanted him not to be burdened. He used to wait in the car while I was in the clinic. The waiting room is often very busy so he waited in the car. He could not leave the car in case I finished early. So I decided to apply for the service partly for him. Now it costs me only ¥218 one way – it costs more than ¥900 if I use an ordinary taxi.

(Mrs Gotoh, aged 71, service user)

In societies such as Britain, where concepts of rights are fully developed, people feel – as a result of paying contributions either through taxation and/or national insurance – that they have a right to access services to which they have contributed (Plant 2001). For older Japanese people, although rights to public pensions and healthcare have been fully developed, it was initially thought that it could be difficult to remove the stigma previously attached to the old 'public welfare' and to the notion of unfulfilled obligation by the family. For some, there are still difficulties in adjusting to a new social environment to receive formal support, since allowing strangers to come to their house and to provide personal and nursing care through physical contacts would still be a challenge. However, the steady increase in the take-up rates, especially for domiciliary services, for the first five years meant that older people and their families with various needs welcomed the scheme. However, take-up has not yet reached the expected figures for the first period. As Hiraoka (2006) indicates, only a limited number of service users fully use their maximum approved amount – the average actual take-up rate was 42 per cent of each maximum approved limits/amounts in 2002. Apart from emotional hesitation, the cost (user fees) and the quality of available services were some of the reasons behind the underuse of approved services (ibid.).

A generation gap may also still exist, since to a certain extent, younger family members who are willing to shift their responsibility to the state have welcomed the system:

Mrs Yamashita had looked after her husband who became disabled after a stroke single-handedly. 'Long term care insurance? I do not want to be

Rethinking the 'generational contract'

looked after by the state. Does not everyone say so? In fact, if you became poorly and unable to think straight, the family is the one who wants to use such a scheme. It is not the person who needs care that decides, but the family.

(Mrs Yamashita, aged 76)

The reason behind the popularity of day services is that it provides family members respite care. There were cases even in the snow, where a daughter-in-law insisted her father-in-law go to a day centre. Even though he was coughing, she put a woolly hat on him and pushed his wheelchair outside to be picked up. Co-resident families are the ones who use the services eagerly. In this sense, the social norm has shifted especially among younger people.

(City official)

In contrast, in Britain, where the financing and delivery of care has been organized very differently from Japan, the policy shift has been directed increasingly towards 'cash and care' (see, for example, Glendinning and Kemp 2006). In the most recent policy reform, cash benefits will replace some of the direct services for social care and older people will enjoy more control if they wish to shop around for the best packages of care themselves.

Contested expectations and outcomes

One of the difficulties of inheritance studies addressed in Chapter 3 is that people's intentions do not necessarily match the inheritance outcomes. In a society like Japan where people are not accustomed to making wills, despite the legal definitions of equal shares, many informal arrangements still supersede such legal definitions. Consequently, many adult children, especially daughters, tend to give up their share of inheritance by signing a legal statement of withdrawal in favour of the 'successor' child. This is largely the case when housing is not considered as a tradable commodity but is occupied by the successor child.

In addition, in both societies, different members of the family expressed different expectations regarding care and inheritance. For example, a son assumed that his sister would take over their parental home, but his sister did not want to live in the house, and the mother wanted to move to a smaller place or even to a nursing home if necessary. Such wishes and opinions were contested and not shared among the family members, especially when they 'do not discuss the matter within the family'.

One Japanese case presents a showcase of the issues that contemporary Japanese society has been facing regarding care and inheritance, in particular regarding the role and status of women in the family, work and the inheritance paradigm. Yoko was married with two grown-up children (both still lived at home) and had been a primary carer to her disabled father for two-and-a-half years before he died – one year prior to the interview in 2005. When her mother died in a car accident a few years previously, she recalled her feeling then that 'all the plans she had in her life were suddenly taken away from her'. She had a highly paid job in a major insurance firm. Her father was a wealthy man but difficult to look after, and no one in the family got along with him. Her siblings (a brother and two sisters) were all unwilling to provide care to their father due to various health and financial reasons, on top of the personality incompatibility. Having felt the obligation as the only son, her younger brother, with his third wife, initially moved in with his father, but the relationship fell apart after six months. The siblings then got together and discussed alternatives:

> Because no one wanted to take father in, they tried to persuade me to do it. 'If you do not do anything extravagant, you can live on your husband's salary', they said. I replied, '... are you going to give me some sort of wages? Can you each give me ¥30,000 [£150] every month? You won't, will you?' I said. We still have a mortgage in our house. My brother also quietly said 'If you look after father, you can inherit everything when he dies'. We did not know how much father was worth, and I did not like such an unwritten promise. We may not struggle to eat but what about my career, which I have invested in. I can receive salary at the end of the month; I have a job status that people recognize. If I quit, I will become just a middle-aged woman – no qualification, nothing!'
>
> (Yoko, aged 51)

Yoko eventually quit her job and became a full-time carer. The loss of her opportunity cost (income through paid work) was ignored when the siblings got together again to discuss the asset distribution after his death. This highlights the issue of how carers and care work are in general undervalued in post-industrial societies with developed welfare programmes. And how this woman's position was still defined as an 'additional' earner in the family and society despite having a high income:

> My brother felt he had already given up his inheritance when his co-residency fell out. He suggested I inherit everything. After the funeral, he said 'I would like to give everything to Yoko who quit her job to look after father. Imagine how much she would have earned if she had stayed in her job'... But my sisters disagreed. They twisted around and said 'Yoko would not like that. She may feel a burden if we asked her to take all ... We also supported him emotionally – talked to him on the phone when he had arguments with Yoko'. I did not want to argue any further and proposed to divide up equally. If father had debts, I thought, I could not pay them all myself.
>
> (Yoko, aged 51)

However, she is still bitter about the unfair treatment she received from her sisters. Despite the unfulfilled obligation on her brother's part, her father was a traditional man and wanted to leave everything to his son who would carry his name. This was apparent (although without a will) since the father left savings of up to ¥10 million (£50,000) in her brother's name, and the ownership of his additional properties had already been transferred to the brother. Her brother, however, refused to inherit all the assets intact and suggested owning the residential properties jointly among the siblings. The father had not shown any gratitude to Yoko, who had taken the responsibility to be a full-time, sole carer for two-and-a-half years. In his view, she was still considered to be married off to another family. So not only was her opportunity cost (loss of earnings) ignored but she also bore some of the costs of care herself:

> I did not receive any more than the others, instead I paid in a lot to look after him. My father had money but did not want to give me any. I paid all the time when I took him to the hospital, and went to do some shopping. When I asked for money, he gave me ¥1,000 [£5] but what can you buy with ¥1,000? I did not want him to say that he was supporting my family so I stopped asking for it … He did not leave a will. If he did, I would not have received anything. He wanted to leave everything to my brother and his children [grandsons] from his first marriage because they are boys and lived together until the older one was five years old. He even had savings with the grandson's name. He left nice watches to them as well. But there was nothing for me, anywhere.
>
> (Yoko, aged 51)

Conclusion

Yoko's case highlighted the absence of common values within the family on support and inheritance and also the existing mismatch of traditional and contemporary values. It also highlighted the varying expectations of different members of the family over their role within and contributions towards the family as well as over their rights to the share of assets. Despite the legal definition, gender bias is still strongly held by many people over care provision and inheritance practices, especially by the older generation. In Britain, in contrast, the majority of the informants exhibited a strong consensus over inheritance – the equal share and treatment of children regardless of their contributions. Moreover, the owners of assets often controlled their disposal through making a will, while in Japan such practice has not yet found a firm root and can thus lead to conflict, dissatisfaction and unequal sharing of assets among family members. Yoko's case, however, interestingly illustrated why an equal share of assets among adult children might look unfair under certain circumstances.

Rethinking the 'generational contract'

The British data on attitudes and practice towards the exchange of inheritance and family support were more predictable and exhibited a greater degree of consensus among the informants, both young and old. In contrast, Japanese responses were more divided within and between the generations. In Japan, the generational contract is no longer very clear-cut, and the contradictions and ambiguities of attitudes and practices exhibited by the current generations clearly explain this recent social transition. Overall, the research found no strong link between bequest motives and family support provision for long-term care in both societies, which had previously been considered as a vital exchange tradition, at least in Japan. Today the new social insurance in Japan has made it possible for older people to live independently, even within co-residency. In this policy context, the relationship between care and inheritance has been increasingly grown apart. In general, housing assets were not used to negotiate family support and common and shared responses among older people were 'not to rely on their children in old age but willing to leave them something'. 'Wishing the assets to be used by the older parents themselves' was a commonly expressed view by adult children in both Britain and Japan. Different scenarios involving caring older parents have produced different and conditional responses, illustrating the complex and dynamic nature of the caring relationships in the two societies.

8 Conclusion

Convergence and diversity

Trends of policy convergence and diversity across societies are often the central concern of comparative social welfare research. As Kennett (2001) states, early work on the growth of welfare states emphasized the growing convergence in social policy development among all the nations of the older industrialized world. For example, it is argued that some levels of economic development often serve as a prerequisite for effective welfare programmes and thus the state adopts welfare policies at different levels of industrialization. Particular stages of economic and technological growth and industrialization thus tend to encourage convergent welfare state forms despite the differences in cultures, political ideologies and national histories (see, for example, Wilensky 1975). More recently, some scholars have argued that factors such as participation in a tight-knit global policy-making community, the cross-national emulation of welfare policies, and the narrow range of policy options available in the fields may stimulate the adoption of welfare policies and thus encourage convergence in an increasingly globalizing world (Kasza 2006). In this context, using the case of Japan's health and pension policies, Kasza (ibid.) highlights how Japan's programmes resemble those of Western Europe more closely today than in the past. In particular, in the post-bubble era, the distinctive features of Japanese social policy are either disappearing (for example, a decline in strong occupational welfare provision) or spreading out to other countries (for example, social insurance schemes on long-term care being adopted in other East Asian societies such as Korea). Such developments tend to support the convergence thesis (Izuhara 2003a; Kasza 2006).

The dominant use of aggregated data in comparative policy analysis has also been critiqued as part of the reason for the early popularity of the convergence thesis. It has indeed been established that different approaches and analyses could lead to rather different conclusions in comparative social research (see, for example, Hill 2006). As Taylor-Gooby (2002) suggests, while quantitative analysis tends to

Conclusion

overemphasize continuity and stability, case studies often present more complexity, diversity and differences in nations' social and political processes and outcomes, and also highlight the change and the instability of current settlements. With the gradual inclusion of qualitative evidence, academic debates in this field have moved on to recognize greater diversity among different groups of people, and cultural and institutional differences in different societies and regions (see, for example, Sainsbury 1994). Moreover, for the past two decades, regime theory, clustering 'families of nations' with similar characteristics, has also become one of the dominant comparative frameworks and has generated a significant amount of academic debates in the field of comparative social policy (for example, Esping-Andersen 1990, 1996; Bonoli 2000; Alcock and Craig 2001; Hill 2006; Takegawa and Lee 2006). Indeed, work on East Asia has contributed to the further development of welfare regime theories. Features such as the distinctive ways in which East Asian societies have organized their welfare policy, striking differences within the region as well as its close relationship between economy and welfare, have fed into the debates around the 'fitness' of the regime clusters. Looking at the East is now an integral part of comparative social and policy analysis.

One of the key policy areas explored in this book is long-term care for older people. Substantial convergence has been observed in this policy arena between Britain and Japan as well as globally but with some distinctive differences. First, the necessity to have policy in this area is shared by many highly developed economies, which are also rapidly ageing societies. In Japan as well as in other East Asian societies, this has been a long-neglected area of social policy due to the traditional reliance on families in this particular welfare field. Given the inevitable link between social policy and economic development in those 'developmental states', the upward convergence trajectory has been witnessed among particular East Asian societies but not across the board in the region. For example, Korea and Taiwan have been 'catching up' Japan in terms of social policy development, while other states, including Singapore and Hong Kong, remain 'residual' in their state involvement in this particular policy field. The trajectories of social policy development are therefore rather polarized, particularly after the Asian Financial Crisis in the region.

Among developed Western welfare states, converging trends regarding long-term care provision in recent years include home-centred care; a mixed economy approach with increased services and providers; and also increasing care options including cash benefits to informal carers. However, the process to achieve such policy goals and its impacts on different groups are subject to substantial national variation. For example, a trend towards home care, or the promotion of 'ageing in place', may mean de-institutionalization in some societies including Britain, but for others (for example, Japan) it is by-passing a development stage of supplying a sufficient number and quality of institutions.

Conclusion

The diversity also exists in other areas such as eligibility criteria (for example, universal versus means-tested), mechanism of funding and delivery (for example, tax-based scheme versus social insurance), and levels and types of benefits and services available. In this context, Japan has opted for a social insurance scheme which exhibits policy convergence towards many established Western European welfare states such as Germany. At the same time, the Japanese policy is attracting other East Asian societies (notably Korea and Taiwan) to follow suit. The restriction of care options, such as not allowing cash benefits to informal carers, is Japan's distinctive (non-convergent) feature, while this option is increasingly adopted in many welfare states regardless of their funding or delivery mechanisms. The relationship between the family, the state, and welfare is perhaps something that still needs to be negotiated in the East Asian welfare mix.

How the state perceives individuals' (housing) asset accumulation in relation to welfare was another policy concern in this book. Policy convergence (which occurred much earlier in postwar history) towards the promotion of home ownership has been apparent between Britain and Japan, with again their own distinctive processes. Whereas in Japan the direct provision of low-interest loans by the state agency as well as strong company housing schemes supported the expansion process, in Britain the sale of council housing since the 1980s has significantly boosted the rates of home ownership and helped shift the profile of the sector. The latter approach was not available in Japan where the provision of public housing has always been minimal. Divergence trends have arisen more recently from the contrasting economic fortunes of the two societies and the different housing markets that thus have impacted on people's ideology and aspirations associated with home ownership. Although the shared policy direction has created strong consensus on the value of home ownership in the two societies, the recent contrasting experiences have weakened such consensus, especially among the younger age cohorts in Japan. In this context, expressing positive choices towards private renting by younger Japanese people in the rather stagnant housing market appears to be somewhat different from the struggle of prospective younger buyers not being able to get onto the property ladder in the inflated housing market in Britain. Such Japanese attitudes may be rectified with the recovery of the economy but this study has certainly suggested a shift away from the conventional notion of home ownership as the most desirable tenure choice at the turn of the century.

In policy terms, the link between care provision and inheritance is defined rather differently in the two societies. The asset-based welfare approach is more evident in England (Scotland has a different policy) where older people are expected to use their accumulated assets including their residential property to meet their care needs in old age. While in Japan, such a (housing asset-based) approach is still new and thus attitudes of individuals, families as well as policy

Conclusion

makers are not well formed or fixed. Expectations of using housing assets in old age are not at the forefront of policy direction in Japan. Instead, the transfer of assets over generations is strongly encouraged, especially for housing use through generous *inter-vivo* allowances, but this is perhaps largely for stimulating the current depressed economy. The separation between care provision and asset transfer is now evident, and the current system encourages the further accumulation of trans-generational wealth. In terms of inheritance tax, Britain has recently doubled the tax-free threshold for married couples (up to £600,000 in 2007). In Japan, there is a contrasting academic debate about re-inventing the notion of the 'social contract' to fill the gap in the micro-level 'generational contract'. Atsumi (2001) argues that if exchanging care and inheritance, which once formed a core of the generational contract, is no longer practised in the family, and also if the provision of social care is no longer means tested, inheritance tax may be used more effectively to redistribute individual wealth back to society. The need to shift the function of the micro-level 'generational contract' to a 'social contract' may be increasingly highlighted in Japan's welfare debate. Overall, such policy differences tend to bring different outcomes and help form different attitudes in society towards housing, care and inheritance dynamics.

Analysing how and why Japan is converging on UK family norms and practices is another important point, considering in particular the relationship between social policy and social change. Have social policy and social change brought the generations together or helped widen the gap between them? Are there significant cross-national differences in terms of distinctive cohort effects by the nations' social policy and social change?

There are clear cohort effects that divide winners and losers in the nations' social, economic and policy contexts. This study focused on 'generations' in terms of relationships within families rather than 'cohorts' – age groups regardless of family relations. It is interesting, however, to examine who wins what aspects at the combination of changes in the micro-level interpersonal relations and more macro social context. Considering macro contexts such as changes in the economy, housing markets and the development of social security, significant differences exist in Japan between baby-boomers (born between 1946 and 1950 in Japan) and the subsequent cohort due to their timing of entry to home ownership, their experience of economic recession, and thus the experience of capital gains and losses (Hirayama 2006). The impact of the 'lost decade' of the 1990s appears to have placed a heavy burden on the post-baby-boomer cohorts. Such economic factors tend to influence people's behaviour with regard to family formation and family relations. Who gained in the context of an ageing society with developed social security benefits and services? The newly introduced social insurance scheme in Japan has significantly lifted a burden of family care from adult children; and children are now more likely to receive a larger inheritance from their parents due to the increased wealth accumulated by the

Conclusion

parent generation and also fewer children born in families. Collectively, however, the social contract now places greater financial responsibility in the current and subsequent working-age cohorts despite increasing insecurity in the labour market.

Societal ageing is a shared demographic trend in the global North but the profound speed of ageing found in more recently industrialized societies in East Asia has its own significant implications on intergenerational relations. With globalization, social change and economic development, in general, family ideology and practices are indeed converging across the two societies. In Britain, this study found strong consensus over the generations regarding many aspects of inheritance practices, including fairness and equality attached to adult child beneficiaries. In Japan, on the other hand, more profound social, demographic changes often meant a widening generational gap between current cohorts of the young and the old, especially between conventional ideas and actual patterns of family support; as well as between the legal definition and actual practices of inheritance. The generations adopt these changes but the compressed social change in Japan tends to alter social norms, expectations and cultural practices and create more conflict and ambivalence among the generations. Such consensus and conflict models clearly highlighted the East–West differences.

In terms of the particular 'generational contract' exchanging care and inheritance, as identified in Chapter 7, there are a number of reasons why the Japanese practice is moving away from their conventional arrangement embedded in the traditional family system. Institutional change accompanying social change is one strong driving force; however, changes in laws and policy do not necessarily always bring changes in practices spontaneously. The policy/legal intention of democratizing families, whether or not influenced by the West, has been slow to materialize in some areas partly due to the strong traditions and socioeconomic structure of the nation. For example, although it has been more than 50 years since the postwar inheritance law first defined equal treatment of children as beneficiaries, regardless of gender or birth order, full practice is still slow to come by as mixed views and practices have been observed in different families in Japan. It was evident that many informal arrangements of passing housing assets intact to one particular child could still supersede the legal definition of the equal treatment. The strong resistance of the East to 'converge' to the West can be explained by the existence of different 'cultures' here – symbolized around traditional living arrangement of co-residence, the importance of family continuity and ancestor worship and so on. The boundary regarding care responsibility between the family and the state has shifted significantly, however, and under the new social insurance scheme, caring responsibility is shared by a variety of providers in Japan that resembles UK policy and practice. The development of social security, housing policy centring on home ownership, and occupational welfare provision has also influenced people's attitudes and choices

Conclusion

regarding family practices. This policy change has indeed contributed to the separation of the previously close-knit 'generational contract' between care and inheritance. At the same time, wider social change, such as changing employment patterns and participation, increased geographic mobility and declining family size all appear to be equally convincing explanations for changing family practices. Policy can, however, be both a cause and a consequence of social change, as the social insurance on long-term care introduced to fill the care gap already vacated by families inevitably helps further accelerate changes taking place in Japanese families.

Finally, distinctive contrasts in inheritance practices between Britain and Japan are largely gender-related. The fact that women tend to shoulder a heavier burden of care-giving is shared but the disadvantageous position of women in asset building and inheritance appeared to be more Asian/Japanese. Moreover, while many Japanese practices are converging to those of the British, a recent increase in 'complex families' in Britain means a departure from their conventional – simpler and rather predictable – bequest attitudes and plans. Increasing rates of divorce and re-partnering complicate existing family structures, function and relations that create a series of individualized relationships within the family, and impact on the previously defined 'generational contract' regarding caring responsibilities and the division of assets among children. The dynamic nature of families and social relations will therefore continue to provide topics of inquiry for comparative social research.

References

Ackers, L. (1998) *Shifting Spaces: Women, citizenship and migration within the European Union*, Bristol: The Policy Press.
Ackers, L. (2004) 'Citizenship, migration and the valuation of care in the European Union', *Journal of Ethnic and Migration Studies*, 30(2): 373–96.
Akiyama, H., Antonucci, T.C. and Campbell, R. (1997) 'Exchange and reciprocity among two generations of Japanese and American women', in J. Sokolovsky (ed.) *The Cultural Context of Aging: Worldwide perspectives*, 2nd edn, Westport, CT: Bergin & Garvey: 163–78.
Alcock, P. and Craig, G. (eds) (2001) *International Social Policy: Welfare regimes in the developed world*, Basingstoke: Palgrave.
Allan, G. (1982) 'Property and family solidarity', in P. Hollowell (ed.) *Property and Social Relations*, London: Heinemann.
Arakawa, T. (2003a) 'Koreisha hoyu shisan no genjo to sozoku' ['Older people's assets and inheritance'], *Life Design Report*, 2003(5): 16–23.
Arakawa, T. (2003b) 'Koreisha shisan no ryudo-ka' ['Liquidising assets among older people'], *Life Design Report*, 2003(9): 4–15.
Arber, S. and Attias-Donfut, C. (eds) (1999) *The Myth of Generational Conflict: The family and state in ageing societies*, London: Routledge.
Arksey, H. and Glendinning, C. (2007) Choice in the context of informal care–giving, *Health and Social Care in the Community*, 15(2): 165–75.
Atoh, M., Kandiah, K. and Ivanov, S. (2004) 'The second demographic transition in Asia?: Comparative analysis of the low fertility situation in East and South-East Asian countries', *The Japanese Journal of Population*, 2(1): 42–75.
Atsumi, N. (2001) 'Shakai hosho zaigen toshiteno sozoku shisan no katsuyo' ['Utilization of inheritance assets as social security resources'], *Fuji Research Institute Corporation Study Report*, December 2001.
Attias-Donfut, C. and Wolff, F.C. (2000) 'The redistributive effects of generational transfers', in S. Arber and C. Attias-Donfut (eds) *The Myth of Generational Conflict: The family and state in ageing societies*, London: Routledge: 22–46.
Barnes, M. (2006) *Caring and Social Justice*, Basingstoke: Palgrave Macmillan.

References

Becker, H. (2000) 'Discontinuous change and generational contracts', in S. Arber and C. Attias-Donfut (eds) *The Myth of Generational Conflict: The family and state in ageing societies*, London: Routledge: 114–32.

Beltran, A. (2000) *Grandparent and Other Relatives Raising Children: Supportive public policies*, Public Policy and Aging Report 11(1), Washington: Policy Institute of the Gerontological Society of America.

Bengtson, V.L. and Schrader, S. (1982) 'Parent–child relations', in D. Mangen and W.A. Peterson (eds) *Research Instruments in Social Gerontology*, vol 2, Minneapolis, MS: University of Minnesota Press: 115–86.

Bengtson, V.L. and Harootyan, R.A. (1994) *Intergenerational Linkages: Hidden connections in American society*, New York: Springer.

Bengtson, V.L. and Giarrusso, R. (2002) 'Solidarity, conflict, and ambivalence: complementary or competing perspectives on intergenerational relationships?', *Journal of Marriage and the Family*, 64(2): 568–76.

Bengtson, V.L., Rosenthal, C.J. and Burton, L.M. (1996) 'Paradoxes of families and aging', in R.H. Binstock and L.K. George (eds) *Handbook of Aging and the Social Sciences*, 4th edn, San Diego, CA: Academic Press: 253–82.

Bettoni, F. (2006) 'Strong in tradition and yet innovative: the puzzles of the Italian family', in M. Rebick and A. Takenaka (eds) *The Changing Japanese Family*, Oxon: Routledge: 54–71.

Bonoli, G. (2000) *European Welfare Futures: Towards a theory of retrenchment*, Cambridge: Polity Press.

Brannen, J., Moss, P. and Mooney, A. (2004) *Working and Caring over the Twentieth Century: Change and continuity in four-generation families*, Basingstoke: Palgrave Macmillan.

Burgoyne, C. and Morison, V. (1997) 'Money in remarriage: keeping things simple – and separate', *Sociological Review*, 45(3): 363–95.

Burton, L. (1992) 'Black grandmothers rearing children of drug-addicted parents', *The Gerontologist*, 32(6): 744–51.

Butlin, N. (1976) *Investments in Australian Economic Development, 1861–1900*, Canberra: Research School of Social Sciences, Australian National University.

Cabinet Office (2001) *Koreisha no keizai seikastu ni kansuru ishiki chosa* [*Opinion Survey of the Economic Life of Older People*], Tokyo: Government of Japan.

Cabinet Office (2002) *Report on the Fifth International Survey of Lifestyles and Attitudes of the Elderly*, Tokyo: Gyosei Printing Company.

Carers UK (2005) *A Manifesto for Carers*, London: Carers UK.

Chamberlayne, P. and King, A. (2000) *Cultures of Care: Biographies of carers in Britain and the two Germanies*, Bristol: The Policy Press.

Chapman, M. and Sinclair, S. (2003) *Equity Shares in Social Housing*, London: Office of the Deputy Prime Minister.

References

Chau, R.C.M. and Yu, W.-K. (2005) 'Is welfare unAsian?', in A. Walker and C.-k. Wong (eds) *East Asian Welfare Regimes in Transition: From Confucianism to globalisation*, Bristol: The Policy Press: 21–45.

Cherlin, A. and Furstenberg, F. (1986) *The New American Grandparent*, Cambridge, MA: Harvard University Press.

Chi, I., Mehta, K.K. and Howe, A.L. (eds) (2001) *Long-Term Care in the 21st Century: Perspectives from around the Asia-Pacific Rim*, New York: The Haworth Press.

Coleman, M., Ganong, L.H., Hans, J.D., Sharp, E.A. and Rothrauff, T.C. (2005) 'Filial obligations in post-divorce stepfamilies', *Journal of Divorce & Remarriage*, 43(3/4): 1–27.

Comas-Herrera, A., Wittenberg, R., Costa-Font, J., Gori, C., Di Maio, A., Patxot, C., Pickard, L., Pozze, A. and Rothgang, H. (2006) 'Future long-term care expenditure in Germany, Spain, Italy and the United Kingdom', *Ageing and Society*, 26(2): 285–302.

Connidis, I.A. (2005) 'Sibling ties across time: the middle and later years', in M.L. Johnson (ed) *The Cambridge Handbook of Age and Ageing*, Cambridge: Cambridge University Press: 429–36.

Connidis, I.A. and McMullin, J.A. (2002) 'Sociological ambivalence and family ties: a critical perspective', *Journal of Marriage and Family*, 64: 558–67.

Council of Mortgage Lenders (2001) *Changing Households, Changing Housing Markets*, July, London: Council of Mortgage Lenders.

Cowgill, D.O. and Holmes, L. (eds) (1972) *Aging and Modernization*, New York: Appleton-Century-Crofts.

Craib, I. (1992) *Modern Social Theory: From Parsons to Habermas*, 2nd edn, London: Harvester Wheatsheaf.

Daatland, S.O. and Herlofson, K. (2003) 'Lost solidarity' or 'changed solidarity': a comparative European view of normative family solidarity, *Ageing and Society*, 23(5), 537–60.

Daniels, N. (1988) *Am I My Parents' Keeper?*, Oxford: Oxford University Press.

Deeming, C. and Keen, J. (2003) 'A fair deal for care in older age? Public attitudes towards the funding of long-term care', *Policy & Politics*, 31(4): 431–46.

Delphy, C. and Leonard, D. (1986) 'Class analysis, gender analysis, and the family', in R. Crompton and M. Mann (eds) *Gender and Stratification*, London: Hutchinson.

Dench, G., Ogg, J. and Thompson, K. (1999) 'The role of grandparents', in R. Jowell, J. Curtice, A. Park and K. Thompson (eds) *British Social Attitudes: The 16th report*, Aldershot: Ashgate.

DETR (Department of the Environment, Transport and the Regions) (2000) *Housing Statistics Summary 004*, March, London: DETR.

DETR (2001) *Housing in England 1999/2000*, London: DETR.

References

DH (Department of Health) (2005) *Carers and Disabled Children Act 2000 and Carers (Equal Opportunities) Act 2004: Combined policy guidance*, London: DH.

Dimmock, B., Bornat, J., Peace, S. and Jones, D. (2004) 'Intergenerational relationships among stepfamilies in the UK', in S. Harper (ed.) *Families in Ageing Societies: A multi-disciplinary approach*, Oxford: Oxford University Press: 82–94.

Doling, J. (1997) *Comparative Housing Policy: Government and housing in advanced industrialized societies*, Basingstoke: Macmillan.

Dowd, J. (1975) 'Aging as exchange: a preface to theory', *Journal of Gerontology*, 30(5): 584–94.

Dupuis, A. and Thorns, D.C. (1996) 'Meanings of home for older home owners', *Housing Studies*, 11(4): 485–501.

English, J. (1979) 'What do grown children owe their parents?', in O. O'Neill and W. Ruddick (eds) *Having Children: Philosophical and legal reflections on parenthood*, New York: Oxford University Press: 351–6.

Esping-Andersen, G. (1990) *The Three Worlds of Welfare Capitalism*, Cambridge: Polity Press.

Esping-Andersen, G. (ed) (1996) *Welfare States in Transition: National adaptations in global economies*, London: Sage.

Evers, A., Pijl, M. and Ungerson, C. (eds) (1994) *Payments for Care: A comparative overview*, Aldershot: Avebury.

Finch, J. (1989) *Family Obligations and Social Change*, Cambridge: Polity Press.

Finch, J. (1997) 'Individuality and adaptability in English kinship', in M. Gullestad and M. Segalen (eds) *Family and Kinship in Europe*, London: Pinter: 129–45.

Finch, J. (2004) 'Inheritance and intergenerational relations', in S. Harper (ed.) *Families in Ageing Societies: A multi-disciplinary approach*, Oxford: Oxford University Press: 164–75.

Finch, J. and Mason, J. (1993) *Negotiating Family Responsibilities*, London: Routledge.

Finch, J. and Hayes, L. (1994) 'Inheritance, death and the concept of the home', *Sociology*, 28(2): 417–33.

Finch, J. and Wallis, L. (1994) 'Inheritance, care bargains, and elderly people's relationships with their children', in D. Challis and B.P. Davies (eds) *Health and Community Care: UK and international perspectives*, Aldershot: Gower.

Finch, J. and Mason, J. (2000) *Passing On: Kinship and inheritance in England*, London: Routledge.

Finch, J., Hayes, L., Mason, J., Masson, J. and Wallis, L. (1996) *Wills, Inheritance, and Families*, Oxford: Clarendon Press.

Fine, M. (2007) *A Caring Society? Care and the dilemmas of human service in the 21st century*, Basingstoke: Palgrave Macmillan.

References

Forrest, R. (2005) 'Globalisation and the housing asset rich: geographies, demographies and policy convoys', Keynote paper, Asia-Pacific Network for Housing Research (APNHR) Conference on 'Globalisation and Housing', University of Kobe, Japan, September.

Forrest, R. and Williams, P. (1984) 'Commodification and housing: emerging issues and contradictions', *Environment and Planning A*, 16: 1163–80.

Forrest, R. and Murie, A. (1989) 'Differential accumulation: wealth, inheritance and housing policy reconsidered', *Policy & Politics*, 17(1): 25–41.

Forrest, R. and Murie, A. (eds) (1995) *Housing and Family Wealth: Comparative international perspectives*, London: Routledge.

Forrest, R. and Lee, J. (eds) (2003) *Housing and Social Change: East-West perspectives*, London: Routledge.

Forrest, R. and Lee, J. (2004) 'Cohort effects, differential accumulation and Hong Kong's volatile housing market', *Urban Studies*, 41: 2181–96.

Forrest, R., Kennett, P. and Leather, P. (1999) *Home Ownership in Crisis? The British experience of negative equity*, Aldershot: Ashgate.

Forrest, R., Kennett, P. and Izuhara, M. (2003) 'Home ownership and economic change in Japan', *Housing Studies*, 18(3): 277–93.

Forrest, R., Murie, A., Hawes, D., Bridge, G. and Smart, G. (1995) *Leaseholders and Service Charges in Former Local Authority Flats*, London: HMSO.

Fox Harding, L. (1996) *Family, State and Social Policy*, Basingstoke: Macmillan.

Gale, W.G. and Scholz, J.K. (1994) 'Intergenerational transfers and the accumulation of wealth', *Journal of Economic Perspectives*, 8(4): 145–60.

Ganong, L.H. and Coleman, M. (1998) 'Attitudes regarding filial responsibilities to help elderly divorced parents and stepparents', *Journal of Aging Studies*, 12(3): 271–90.

George, L.K. (1986) 'Caregiver burden: Conflict between norms of reciprocity and solidarity', in K. Pillemer and R. Wolf (eds) *Elder Abuse: Conflict in the Family*, Dover, MA: Auburn House.

Giarrusso, R., Silverstein, M., Gans, D. and Bengtson, V.L. (2005) 'Aging parents and adult children: new perspectives on intergenerational relationships', in M. Johnson (ed.) *The Cambridge Handbook on Age and Ageing*, Cambridge: Cambridge University Press: 413–21.

Giddens, A. (1991) *Modernity and Self Identity*, Oxford: Polity Press.

Glendinning, C. and Kemp, P.A. (eds) (2006) *Cash and Care: Policy challenges in the welfare state*, Bristol: The Policy Press.

Goode, W.J. (1963) *World Revolution and Family Patterns*, New York: Free Press.

Gouldner, A.W. (1960) 'The norm of reciprocity: a preliminary statement', *American Sociological Review*, 25(2): 161–78.

Groves, R., Murie, A. and Watson, C. (eds) (2007) *Housing and the New Welfare State: Perspectives from East Asia and Europe*, Aldershot: Ashgate.

References

Gurney, C. (1990) *The Meaning of Home in the Decade of Owner Occupation: Towards an experiential research agenda*, Working Paper 88, Bristol: SAUS Publications.

Hall, R. (1988) *Enterprise Welfare in Japan: Its development and role*, Discussion Paper, Suntory-Toyota International Centre for Economics and Related Disciplines, June.

Hamnett, C. (1999) *Winners and Losers: Home ownership in modern Britain*, London: UCL Press.

Hamnett, C., Harmer, M. and Williams, P. (1991) *Safe As Houses: Housing inheritance in Britain*, London: Paul Chapman.

Hann, C.M. (ed.) (1998) *Property Relations: Reviewing the anthropological tradition*, Cambridge: Cambridge University Press.

Harper, S. (2004) 'Grandparenthood', in M. Johnson (ed.) *The Cambridge Handbook on Age and Ageing*, Cambridge: Cambridge University Press: 422–8.

Harper, S. (2005) 'Migrant female care labour and its impact on family networks', Paper presented at the British Society of Gerontology Annual Conference 14–16 July 2005, Keele.

Harper, S., Smith, T., Lechtman, Z., Ruchiva, I. and Zeilig, H. (2004) *Grandmother care in lone parent families*, Research Report, Oxford: Oxford Institute of Ageing.

Hashimoto, A. (1993) 'Family relations in later life: A cross-cultural perspective', *Generations*, Winter: 24–6.

Hashimoto, A. (2004) 'Culture, power, and discourse of filial piety in Japan: the disempowerment of youth and its social consequences', in C. Ikels (ed.) *Filial Piety: Practice and discourse in contemporary East Asia*, Stanford, CA: Stanford University Press: 182–97.

Hashimoto, A. and Kendig, H.L. (1992) 'Aging in international perspective', in H.L. Kendig, A. Hashimoto and L.C. Coppard (eds) *Family Support for the Elderly: The international experience*, Oxford: Oxford University Press.

Heywood, F., Oldman, C. and Means, R. (2002) *Housing and Home in Later Life*, Buckingham: Open University Press.

Hill, M. (1996) *Social Policy: A comparative analysis*, London: Prentice-Hall.

Hill, M. (2006) *Social Policy in the Modern World: A comparative text*, Oxford: Blackwell.

Hills, J. (1996) 'Does Britain have a welfare generation?', in A. Walker (ed.) *The New Generational Contract: Intergenerational relations, old age and welfare*, London: UCL Press: 56–80.

Hiraoka, K. (2002) 'Kaigo service shijo no jokyo' ['The current state of the care service market'], in *Kaigo service kyokyu system no sai-hensei no seika ni kansuru kyoka kenkyu* [*Report for the Ministry of Health, Labour and Welfare Research Grant: 56–76*].

References

Hiraoka, K. (2006) 'Kironi Tatsu Nihon no Kaigo Hoken' ['Long-term care insurance in Japan at a turning point'], in S. Takegawa and H.-K. Lee (eds) *Fukushi Reji-mu no Nikkan Hikaku: Shakai Hosho, Jyenda, Rodo Sijyo [Comparing Japanese and Korean Welfare Regimes: Social security, gender and labour markets]*, Tokyo: Tokyo University Press: 123–45 [in Japanese].

Hirayama, Y. (2003a) 'Home-ownership in an unstable world: the case of Japan', in R. Forrest and J. Lee (eds) *Housing and Social Change: East-West perspectives*, London: Routledge: 140–61.

Hirayama, Y. (2003b) 'Housing policy and social inequality in Japan', in M. Izuhara (ed.) *Comparing Social Policies: Exploring new perspectives in Britain and Japan*, Bristol: The Policy Press: 151–71.

Hirayama, Y. (2006) *Tokyo no Hate ni [Beyond Tokyo]*, Tokyo: NTT Publications.

Hirayama, Y. and Izuhara, M. (*forthcoming* 2008) 'Women and housing assets in the context of Japan's home-owning society', *Journal of Social Policy*, 37(4): 1–20.

Hirayama, Y., Forrest, R., Hinokidani, M., Izuhara, M. and Kennett, P. (2003) 'Restructuring of the home ownership systems in Japan and Britain', *Housing Research Foundation Annual Report*, 29: 229–40.

Hirose, S., Mifune, M. and Uemura, K. (1998) *Zaisan, kyodo-sei, gender [Assets, partnership and gender]*, Tokyo: Tokyo Women's Foundation.

Holmans, A.E. and Frosztega, M. (1994) *House Property and Inheritance in the UK*, London: Department of the Environment.

Holliday, I. (2000) 'Productivist welfare capitalism: social policy in East Asia', *Political Studies*, 48: 706–23.

Homans, G.C. (1942) *English Villagers of the Thirteenth Century*, Cambridge, MA: Harvard University Press.

Horioka, C.Y., Fujisaki, H., Watanabe, W. and Ishibashi, N. (1998) 'Chochiku doki ian doki no nichibei hikaku' ['A US-Japan comparison of saving and bequest motives'], in C.Y. Horioka, and K. Hamada, (eds) *Nichibei Kakei no Chochiku Kodo [The saving Behaviour of US and Japanese Household]*, Tokyo: Nihon Hyoronsha.

Howe, A.L. (2001) 'Lessons learned, questions raised', in I. Chi, K.K. Mehta and A.L. Howe (eds) *Long-Term Care in the 21st Century: Perspectives from around the Asia-Pacific Rim*, New York: The Haworth Press: 233–41.

Ikegami, N. and Campbell, J.C. (2002) 'Choices, policy logics and problems in the design of long-term care systems', *Social Policy and Administration*, 36(7): 719–34.

Ikels, C. (ed.) (2004) *Filial Piety: Practice and discourse in contemporary East Asia*, Stanford, CA: Stanford University Press.

Institute for Research on Household Economics (2006) *Josei no Life Course to Jutaku Shoyu [Women's Life-course and Home Ownership]*, Tokyo: Satoh Insatsu.

References

International Longevity Centre (2003) *The Giving Age: Inheritance in the context of an ageing population*, Report prepared by the Future Foundation.

Iwasawa, M. (2001) 'Kekkon shinai koibito tachi' ['Couples not willing to marry'], in S. Kawatomo (ed.) *Ronso Shoshika Nihon [Debate: Low fertility Japan]*, Tokyo: Chuo Koron Sha: 51–69.

Izuhara, M. (2000) *Family Change and Housing in Post-War Japanese Society: The experiences of older women*, Aldershot: Ashgate.

Izuhara, M. (2002) 'Care and inheritance: Japanese and English perspectives on the "generational contract"', *Ageing and Society*, 22(1): 61–77.

Izuhara, M. (ed.) (2003a) *Comparing Social Policies: Exploring new perspectives in Britain and Japan*, Bristol: The Policy Press.

Izuhara, M. (2003b) 'Social inequality under a new social contract: long-term care in Japan', *Social Policy & Administration*, 37(4): 395–410.

Izuhara, M. (2004) 'Negotiating family support? The "generational contract" between long-term care and inheritance', *Journal of Social Policy*, 33(4): 649–65.

Izuhara, M. (2005) 'Residential property, cultural practices and the "generational contract" in England and Japan', *International Journal of Urban and Regional Research*, 29(2): 327–40.

Izuhara, M. (2007) 'Turning stock into cash flow: strategies using housing assets in an ageing society', in Y. Hirayama and R. Ronald (eds) *Housing and Social Transition in Japan*, Oxon: Routledge: 94–113.

Izuhara, M. and Shibata, H. (2002) 'Breaking the generational contract?: Japanese migration and old-age care in Britain', in D. Bryceson and U. Vuorela (eds) *The Transnational Family: New European frontiers and global networks*, Oxford: Berg: 155–69.

Izuhara, M. and Kennett, P. (2006) *Women and Material Assets in Britain and Japan*, Tokyo: Institute for Research on Household Economics.

Janelli, R. and Janelli, D.Y. (1997) 'The mutual constitution of Confucianism and capitalism in South Korea', in R. Brook and H.V. Luong (eds) *Culture and Economy: The shaping of capitalism in Eastern Asia*, Ann Arbor, MI: University of Michigan Press: 107–24.

Johnson, C.L. (1983) 'A cultural analysis of the grandmother', *Research on Aging*, 5(4): 547–67.

Kasza, G.J. (2006) *One World of Welfare: Japan in comparative perspective*, New York: Cornell University Press.

Kawatomo, S. (ed.) (2001) *Ronso Shoshika Nihon [Debate: Low fertility Japan]*, Tokyo: Chuo Koron Sha.

Kennett, P. (2001) *Comparative Social Policy: Theory and research*, Buckingham: Open University Press.

Keyes, C.F. (1983) 'Merit-transference in the Karmic theory of popular Theravada Buddhism', in C.F. Keyes and E.V. Daniel (eds) *Karma: An anthropological inquiry*, Berkeley, CA: University of California Press: 261–86.

References

Kim, I.K. and Kim, C.S. (2003) 'Patterns of family support and the quality of life of the elderly', *Social Indicators Research*, 62(1–3): 437–54.

Koike, N. (2003) An *Opinion Survey on Citizens Focusing on Housing Inheritance in Metropolitan Region*, Kiban Seibi Kodan Sogo Kenkyusho Report No. 134: 26–31.

Künemund, H. and Rein, M. (1999) 'There is more to receiving than needing: theoretical arguments and empirical explorations of crowding in and crowding out', *Ageing and Society*, 19: 93–121.

Land, H. (2005) 'Grandmothers: an undervalued resource or a burden on the younger generations?', paper presented at the 'Childhoods' Conference, Oslo, June.

Latham, M. (2001) 'Stakeholder welfare: an asset-based approach to Australian welfare', in S. Regan and W. Paxton, *Asset-Based Welfare: International experiences*, London: IPPR: 55–73.

Lawton, L., Silverstein, M. and Bengtson, V.L. (1994) 'Solidarity between generations in families', in V.L. Bengtson and R. Harootyan (eds) *Intergenerational Linkages: Hidden connections in American society*, New York: Springer.

Leather, P. (1999) *Age File '99*, Kidlington: Anchor Trust.

Lee, J., Forrest, R. and Tam, W.K. (2003) 'Home-ownership in East and South East Asia: market, state and institutions', in R. Forrest and J. Lee (eds) *Housing and Social Change: East-West perspectives*, London: Routledge: 20–45.

Lewis, J. (1992) 'Gender and the development of welfare regimes', *Journal of European Social Policy*, 2(3): 159–73.

Lloyd, L. (2000) 'Caring about carers: only half the picture?', *Critical Social Policy*, 20(1): 136–50.

Lloyd, L. (2003) 'Caring relationships: looking beyond welfare categories of "carers" and "service users"', in K. Stalker (ed.) *Reconceptualising Work with 'Carers': New directions for policy and practice*, London: Jessica Kingsley Publishers.

Lloyd, L. (2006) 'Call us carers: limitations and risks in campaigning for recognition and exclusivity', *Critical Social Policy*, 26(4): 945–60.

Lowenstein, A. (2005) 'Global ageing and challenges to families', in M.L. Johnson (ed.) *The Cambridge Handbook of Age and Ageing*, Cambridge: Cambridge University Press: 403–12.

Lowenstein, A. (2007) 'Solidarity-conflict and ambivalence: testing two conceptual frameworks and their impact on quality of life for older family members', *Journal of Gerontology Social Sciences*, 62B: 100–7.

Lowenstein, A. and Bengtson, V.L. (2003) 'Challenges of global aging to families in the twenty-first century', in V.L. Bengtson and A. Lowenstein (eds) *Global Aging and Challenges to Families*, Hawthorne, NY: Aldine De Gruyter: 371–9.

References

Luescher, K. and Pillemer, K. (1998) 'Intergenerational ambivalence: a new approach to the study of parent–child relations in later life', *Journal of Marriage and the Family*, 60(2): 413–25.
Lye, D.N. (1996) 'Adult child–parent relationships', *Annual Review of Sociology*, 22: 79–102.
Management and Coordination Agency (2003) *Jutaku Tochi Tokei Chosa* [*Housing and Land Survey of Japan*], Tokyo: Bureau of Statistics.
Marinakou, M. (1998) Welfare states in the European periphery: the case of Greece, in R. Sykes and P. Alcock (eds) *Developments in European Social Policy: Convergence and diversity*, Bristol: The Policy Press.
Marshall, V.W., Rosenthal, C. and Dacink, J. (1987) 'Older parents' expectations for filial support', *Social Justice Research*, 1(4): 405–24.
Marshall, V.W., Matthews, S.H. and Rosenthal, C.J. (1993) 'Elusiveness of family life: a challenge for the sociology of aging', in G.L. Maddox and M.P. Lawton (eds) *Annual Review of Gerontology and Geriatrics*, New York: Springer.
Masuda, M. (2002) 'Kazoku kaigo no hyoka to kaigo hoken' ['Evaluation of family care and long-term care insurance'], *Shukan Shakai Hosho*: 2198–202.
Mills, M.B. (1999) *Thai Women in the Global Labor Force: Consuming desires, contested selves*, Chapel Hill, NC: Rutgers University Press.
Ministry of Finance (2001) *The Taxation System in Japan 2001*, Tokyo.
Ministry of Health and Welfare (1941) *Housing Statistics Survey of Large Cities*, Tokyo.
Ministry of Health and Welfare (2000) *White Paper on Health and Welfare*, Tokyo: Gyosei.
Ministry of Health, Labour and Welfare (2003) *Kaigo Hoken no Jisshi Jokyo* [*The Current State of Long-term Care Insurance*], www.mhlw.go.jp/topics/kaigo/kaigi/020904/2-1.html.
Ministry of Internal Affairs and Communications (2003) *Kakei ni okeru kinyu shisan sentaku nado ni kansuru chosa kekka hokoku-sho* [*Survey on the Financial Asset Choice of Households*], Tokyo.
Ministry of Land Infrastructure and Transport (2003) *Jutaku Juyo Jittai Chosa* [*Housing Demand Survey*], Tokyo.
Ministry of Telecommunications (1998) *Kakei ni okeru kinyu shisan sentaku nado ni kansuru chosa hokoku-sho* [*Survey report on household choices of financial assets*].
Minkler, M. and Roe, K. (1996) 'Grandparents as surrogate parents', *Generations*, 20: 34–8.
Mizoguchi, T. (2002) 'Chiho toshi jumin no kyoju kaireki' ['Housing history of residents in a local city'], in Y. Arai, T. Kawaguchi and T. Inoue (eds) *Nihon no Jinko Ido – Life course to chiiki sei* [*Migration in Japan: Life course and locality*], Tokyo: Kokinshoin.

References

Morel, N. (2006) 'Providing coverage against new social risks in Bismarckian welfare states: the case of long-term care', in K. Armingeon and G. Bonoli (eds) *The Politics of Post-Industrial Welfare States: Adapting post-war social policies to new social risks*, Oxon: Routledge: 227–47.

Morikawa, M. (2001) 'Problems and future directions of the long-term care insurance system in Japan', *Hallym International Journal of Aging*, 3(2): 181–90.

Motel-Klingebiel, A., Tesch-Roemer, C. and Von Kondratowitz, H.-J. (2005) 'Welfare states do not crowd out the family: evidence for mixed responsibility from comparative analyses', *Ageing and Society*, 25(6): 863–82.

Mullins, P. (2000) 'The line of descent in the intergenerational transmission of domestic property', *Housing Studies*, 15(5): 683–98.

Munro, M. (1988) 'Housing wealth and inheritance', *Journal of Social Policy*, 17(4): 417–36.

Munro, M. and Leather, P. (2000) 'Nest-building or investing in the future? Owner-occupiers' home improvement behaviour', *Policy & Politics*, 28(4): 511–26.

Murie, A. and Forrest, R. (1980) 'Wealth, inheritance and housing policy', *Policy & Politics*, 8: 1–19.

Noguchi, Y., Uemura, K., Kitoh, Y. and Midohoka, K. (1988) *Sozoku no Jittai to Eikyo ni Kansuru Chosa Kenkyu [Study on facts and effects of inheritance]*, Tokyo: Keizai Seisaku Kenkyu-sho.

O'Dwyer, L. (1999) 'Housing inheritance and the private rental sector in Australia', *Housing Studies*, 14(6): 755–76.

OECD (Organization for Economic Co-operation and Development) (2005) *Long-term Care for Older People*, Paris: OECD Publishing.

Ogawa, N., Retherford, R.D. and Matsukura, R. (2006) 'Demographics of the Japanese family: entering uncharted territory', in M. Rebick and A. Takenaka (eds) *The Changing Japanese Family*, Oxon: Routledge: 19–38.

Parker, G. and Clarke, H. (2002) 'Making the ends meet: do carers and disabled people have a common agenda?', *Policy & Politics*, 30(3): 347–59.

Paxton, W. (ed.) (2003) *Equal Shares? Building a progressive and coherent asset-based welfare policy*, London: IPPR.

Peng, I. (2002a) 'The new politics of welfare state in developmental context: explaining the 1990s social care expansion in Japan', UNRISD project paper.

Peng, I. (2002b) 'Social care in crisis: gender, demography and welfare state restructuring in Japan', *Social Politics*, 9(3): 411–43.

Phang, S.-Y. (2001) 'Housing policy, wealth formation and the Singapore economy', *Housing Studies*, 16(4): 443–59.

Pierson, P. (2001) 'Coping with permanent austerity: welfare state restructuring in affluent democracies', in P. Pierson (ed.) *The New Politics of the Welfare State*, Oxford: Oxford University Press.

References

Plant, R. (2001) 'Rights and responsibilities in long-term care', in J. Robinson (ed.) *Towards a New Social Compact for Care in Old Age*, London: King's Fund.

Pratt, G. and Hanson, G. (1991) 'On the links between home and work: family-household strategies in a buoyant labour market', *International Journal of Urban and Regional Research*, 15(4): 55–74.

Prendergast, D. (2005) *From Elder to Ancestor: Old age, death and inheritance in modern Korea*, Folkestone: Global Oriental.

Qureshi, H. and Walker, A. (1989) *The Caring Relationships: Elderly people and their families*, Basingstoke: Macmillan.

Regan, S. and Paxton, W. (2001) *Asset-Based Welfare: International experiences*, London: IPPR.

Richards, L. (1990) *Nobody's Home: Dreams and realities in a new suburb*, Melbourne: Oxford University Press.

Roberto, K. and Stroes, J. (1995) 'Grandchildren and grandparents: roles, influences and relationships', in J. Hendricks (ed.) *The Ties of Later Life*, New York: Baywood Publishing Company: 141–53.

Roberts, R.E.L. and Bengtson, V.L. (1990) 'Is intergenerational solidarity a unidimensional construct? A second test of a formal model', *Journal of Gerontology: Social Sciences*, 45: S12–S20.

Ronald, R. (2008) *The Ideology of Home Ownership: Homeowner societies and the role of housing*, Basingstoke: Palgrave Macmillan.

Rowlingson, K. (2000) *Fate, Hope and Insecurity: Future expectations and forward planning*, York: Joseph Rowntree Foundation.

Rowlingson, K. (2004) *Attitudes to Inheritance*, Focus group report, project report, University of Bath (available from the author).

Rowlingson, K. and McKay, S. (2005) *Attitudes to Inheritance in Britain*, Bristol: The Policy Press.

Royal Commission on Long-term Care (1999) *With Respect to Old Age: Long-term care – Rights and responsibilities*, Cm 4192, London: The Stationery Office.

Sainsbury, D. (1994) *Gendering Welfare States*, London: Sage Publications.

Saunders, P. (1990) *A Nation of Home Owners*, London: Unwin Hyman.

Sherraden, M. (1991) *Assets and the Poor: A new American welfare policy*, New York: M.E. Sharpe.

Shin, C.S. and Shaw, I. (2003) 'Social policy in South Korea: cultural and structural factors in the emergence of welfare', *Social Policy & Administration*, 37(4): 328–41.

Silva, E.B. and Smart, C. (eds) (1999) *The New Family?*, London: Sage Publications.

Silverstein, M. and Bengtson, V.L. (1991) 'Do close parent–child relations reduce the mortality risk of older parents?', *Journal of Health and Social Behavior*, 32: 382–95.

Silverstein, M. and Bengtson, V.L. (1994) 'Does intergenerational social support influence the psychological well-being of older parents? The contingencies of declining health and widowhood', *Social Science and Medicine*, 38(7): 943–57.

Silverstein, M., Chen, X. and Heller, K. (1996) 'Too much of a good thing? Intergenerational social support and the psychological well-being of older parents', *Journal of Marriage and the Family*, 58: 970–82.

Sorensen, C. and Kim, S.-C. (2004) 'Filial piety in contemporary urban Southeast Korea: practices and discourses', in C. Ikels (ed.) *Filial Piety: Practice and discourse in contemporary East Asia*, Stanford, CA: Stanford University Press: 153–81.

Spilerman, S. (2000) 'Wealth and stratification processes', *Annual Review of Sociology*, 26: 497–524.

Stryckman, J. and Nahmiash, D. (1994) 'Canada', in A. Evers, M. Pijl and C. Ungerson (eds) *Payments for Care: A comparative overview*, Aldershot: Avebury: 307–19.

Sugden, R. (1992) 'Fairness', in S. Hargreaves Heap, M. Hollis, B. Lyons, R. Sugden and A. Weale (eds) *The Theory of Choice: A critical guide*, Oxford: Blackwell.

Takegawa, S. and Lee, H.-K. (eds) (2006) *Fukushi Regime no Nikkan Hikaku [Welfare Regimes in Japan and Korea: Social Security, Gender and Labour Markets]*, Tokyo: University of Tokyo Press.

Tamai, K. (1997) 'Shakai seisaku kenkyu no keifu to konnichi-teki kadai' ['History of the study of social policy and current issues'], in K. Tamai and M. Ohmori (eds) *Shakai seisaku wo manabu hito no tami ni [For those who study social policy]*, Kyoto: Sekai Shisou sha.

Tao, J. (2004) 'The paradox of care: a Chinese Confucian perspective on long-term care', in P. Kennett (ed.) *A Handbook of Comparative Social Policy*, Cheltenham: Edward Elgar: 131–50.

Taylor-Gooby, P. (2002) The silver age of the welfare state: Perspectives on resilience, *Journal of Social Policy*, 31(4): 597–622.

Theobald, H. (2003) 'Care for the elderly: welfare system, professionalisation and the question of inequality', *International Journal of Sociology and Social Policy*, 23(4/5): 159–85.

Thorns, D.C. (1989) 'The impact of homeownership and capital gains upon class and consumption sectors', *Society and Space*, 7: 293–312.

Thorns, D.C. (1994) 'The role of housing inheritance in selected owner occupied societies (Britain, New Zealand, Canada)', *Housing Studies*, 9(4): 473–92.

Thorns, D.C. (1995) 'Housing wealth and inheritance: the New Zealand experience', in R. Forrest and A. Murie (eds) *Housing and Family Wealth: Comparative international perspectives*, London: Routledge: 8–35.

References

Tobio, C. (2004) 'Kinship support, gender and social policy in France and Spain', in T. Knijn and A. Komter (eds) *Solidarity Between the Sexes and the Generations: Transformations in Europe*, Cheltenham: Edward Elgar.

Tokyo Women's Foundation (1997) *Tsuma to Otto no Zaisan [Assets of Wife and Husband]*, Tokyo: Tokyo Women's Foundation.

Turner, B. (1986) *Equality*, Chichester: Ellis Horwood.

Turner, J.H. (1982) *The Structure of Sociological Theory*, 3rd edn, Chicago, IL: Dorsey Press.

Twigg, J. and Grand, A. (1998) 'Contrasting legal conceptions of family obligation and financial reciprocity in the support of older people: France and England', *Ageing and Society*, 18(2): 131–46.

Uhlenberg, P. (1980) 'Death and the family', *Journal of Family History*, 5(3): 313–20.

Umeda, Y. (2002) 'Intermarriage in rural Japan: a consequence of the changing Japanese family', Unpublished paper presented at the 'Changing Japanese Family in Comparative Perspective' Conference, November, University of Oxford.

Ungerson, C. (1994) 'Morals and politics in "payments for care": an introductory note', in A. Evers, M. Pijl and C. Ungerson (eds) *Payments for Care: A comparative overview*, Aldershot: Avebury: 43–7.

Ungerson, C. (1997) 'Social politics and the commodification of care', *Social Politics*, 4(3): 362–81.

Ungerson, C. (2004) 'Whose empowerment and independence? A cross-national perspective on "cash for care" schemes', *Ageing and Society*, 24: 189–212.

Ungerson, C. and Yeandle, S. (eds) (2007) *Cash-for-care in Developed Welfare States*, Basingstoke: Palgrave Macmillan.

US Census Bureau (1990) *Household Worth and Asset Ownership 1988*, Washington, DC: US Government Printing Office.

US Census Bureau (2001) *An Aging World 2001: International population reports*, Washington DC: US Government Printing Office.

Wade, A. and Smart, C. (2004) Continuity and Change in Parent-Child Relationships over Three Generations, ESRC End of Award Report (R000239523).

Walker, A. (ed.) (1996) *The New Generational Contract: Intergenerational relations, old age and welfare*, London: UCL Press.

Walker, A. and Wong, C.-K. (eds.) (2005) *East Asian Welfare Regimes in Transition: From Confucianism to globalisation*, Bristol: The Policy Press.

Wanless, D. (2006) *Securing Good Care for Older People: Taking a long-term view*, London: King's Fund.

Whyte, M.K. (2004) 'Filial obligations in Chinese families: paradoxes of modernization', in C. Ikels (ed.) *Filial Piety: Practice and discourse in contemporary East Asia*, Stanford, CA: Stanford University Press: 106–27.

References

Wilensky, H.L. (1975) *The Welfare State and Equality: Structural and ideological roots of public expenditures*, Berkeley, CA: University of California Press.

Williams, F. (2004) *Rethinking Families*, London: Calouste Gulbenkian Foundation.

Williams, P. (2003) 'Home ownership and changing housing and mortgage markets', in R. Forrest and J. Lee (eds) *Housing and Social Change: East-West perspectives*, London: Routledge: 162–82.

Yamada, M. (1999) *Parasaito Shinguru no Jidai [The Age of Parasite Single]*, Tokyo: Chikuma Shinsho Press.

Yoneyama, H. (2005) Mansion no shumatsu-ki mondai to aratana kyokyu hoshiki [The issue of the lifespan of condominiums and a new system of provision], Research Report No 239, September, Fujitsu Soken (FRI) Economic Research Institute (www.jp.fujitsu.com/group/fri/downloads/report/research/2005/no 239.pdf).

Yoo, S., Kojima, K., Negami, A. and Uozaki, K. (1999) *A Study of the Stabilization of the Lives of Elderly People by Utilizing their own Houses: an analysis of the reverse mortgage system and its economic effect*, Sumi So Ken Report No. 26: 323–34.

Index

Accumulation of wealth/assests 26, 33, 41, 66, 73, 77, 78, 79
Adachi, Mrs. 106
Africa 23, 24
Ageing population/population ageing 17, 18, 19, 37
Ageing society/societal ageing 2, 3, 18, 19, 21, 37, 50, 55, 111, 132
Ageing in place 36, 56, 60, 132
Akiyama, Mr. 88
Ambivalence/ambivalent 16, 17, 47, 83, 84, 85, 86, 101, 105, 122, 123, 135
    Intergenerational ambivalent model 16, 17
Asian Financial Crisis 26, 31, 132
Assets 4, 11, 22, 27
    Asset-based approach 3, 27, 44, 45, 77, 133
    Asset transfer 28, 29, 37, 40, 112, 134; see also housing assets; family assets
Asymmetrical reciprocity 15
Australia 19, 30, 32, 38, 51, 58, 60, 61
Austria 57, 61
Autonomy 99, 108

Baby-boomer 36, 81, 134
Beanpole family 18
Ben 80, 98, 119
Beneficiaries 37, 38, 39, 93, 102, 122, 135
Bengtson, V. 16, 17, 18, 20
Brown, Mrs. 69, 70, 79
Boom 32, 66, 68, 82
Boundary/boundaries (between the state and the family) 2, 47, 57, 58, 59, 64
Bubble economy 31, 34, 52, 72, 73, 86; see also post-bubble
Buddhist 15, 43
Burden 15, 50, 52, 58, 62, 85, 103, 111, 122, 124, 134, 136

Cabinet Office 40, 89
Capital gain 26, 33, 69, 72, 73, 76, 77, 81, 134
Capital loss 85
Care options 2, 56, 61, 132, 133
Carers UK 119
Caroline 73, 76
Cash benefits 2, 48, 56, 62, 63, 127, 132, 133
Cash for care/cash and care 2, 127
Central Provident Fund 32
China (People's Republic of) 13, 15, 19, 36, 52, 54
Chiyo 84
Civil code 40, 42, 43, 88, 110
Clarke, Mr. and Mrs. 94
Cohort 11, 17, 20, 50, 81, 83, 134
Cohort effects 35, 36, 134
Commodity/commodities 3, 8, 10, 74, 89, 90
    Exchangeable commodities 8, 10, 11, 84
Commodification 2, 63, 89
Commodified 52, 61
Community care 55, 56, 60
Company housing 32, 66
Comparative 2, 4, 5, 7, 27, 30, 48, 50, 131, 132
Condominium 34, 66, 73, 79, 85, 86, 101, 104, 105
Confucian 12, 13, 48, 53, 54
Control 14, 28, 62, 99, 108, 111
Convergence 5, 131, 132, 133, 135
Co-payment 53, 59, 60
Co-residency 15, 20, 30, 36, 40, 51, 53, 68, 82, 85, 89, 90, 91, 103, 110, 112, 113, 115, 123, 124
Cost of care 28, 29, 44, 50, 51, 118
Council housing 26, 27, 30, 31, 68, 133
Cultural logic 16

152

Index

Cultural norm 1, 44
Culture of care 114, 123
Cultural practice 8, 27, 39, 86, 96, 100, 106, 110, 135

Danchi [housing corporation housing] 68, 84
Davies, Mrs. 87, 93, 107
Dawson, Mr. 114
Debts 14, 15, 36, 105
Debts of merit 15
Democracy/democratizing 14, 27, 40, 110, 120, 135
Demographic change 37, 49, 79, 97, 104, 105
Dependency 10, 105
Dependency ratio 19
DETR 33, 97
Developmental state 31, 48, 53, 132
Diversity 16, 71, 131, 132, 133
Divorce 20, 22, 23, 40, 51, 73, 79, 97, 104, 123
Dōkyo see Co-residency
Duffy, Mrs. 69

East Asia 4, 12, 13, 14, 31, 34, 36, 39, 41, 44, 48, 51, 53, 60, 132, 133, 135
Economic and Social Research Council 5
Eligibility 43, 57, 133
Ellie 77, 103, 120
European Union/Europe 2, 18, 19, 22, 30, 48, 52, 60
Evolutionary theory 39
Exchange 1, 8, 9, 10, 11, 12, 120, 130
  Exchange partner 8, 10
  Exchange practice 9, 11, 21
  Exchange rule 10, 12
  Exchange theory 9, 11
  Exchangeability 10, 11
Equal division/divide equally 88, 92, 95, 99, 103, 115, 119

Fairness 23, 39, 93, 94, 95, 97, 111, 114, 135
Familism 2, 13, 22
Familistic 48
Family assets 28, 88, 100; *see also* Assets
Family care 3, 41, 48, 50, 51, 59, 61, 62, 63, 117, 118, 124
Family continuity 13, 28, 74, 75, 88, 89
Family ethics 13, 54
Family ethnography 5

Family obligation 12, 16, 17, 28, 41, 42, 47, 54, 103, 106, 115, 116
Family support 2, 9, 20, 48, 51, 88, 91, 99, 106, 111, 112, 113, 114, 135
Family relations 6, 16, 17, 28, 81, 85, 92, 93, 95, 99
Family responsibility 13
Family solidarity 16, 28, 42, 88, 91, 92, 119
Family wealth 30, 35, 41, 65, 93, 97, 102, 111
Family welfare 81, 93
Fertility (low) 18, 20, 21, 39, 46, 88
  high fertility 67
Filial piety 1, 13, 14, 15, 53
Finch, J. 4, 8, 28, 35, 38, 42, 44, 48, 53, 54, 90, 93, 97, 111
Forrest, R. 5, 11, 26, 27, 29, 34, 35, 36, 39, 81, 84, 90
France 19, 22, 32, 42, 43, 56
Frustrated movers 84
Fujiwara, Mrs. 71, 78

Gaskin, Mrs. 90
Gender 13, 16, 17, 34, 35, 42, 43, 51, 54, 63, 71, 78, 79, 106, 129, 136
Generational bonds 13
Generation(al) gap 17, 126, 135
Generations 1, 5, 8, 12, 20, 30, 35, 36, 134
Generation in the middle/middle generation 23, 51
Generational contract 1, 2, 4, 9, 12, 20, 28, 45, 47, 54, 93, 106, 110, 111, 112, 122, 124, 134, 135, 136
Generational equity 17
Germany 32, 40, 51, 56, 57, 58, 61, 62, 63, 133
Globalization/globalizing world 18, 21, 24, 52, 131, 135
Gotoh, Mrs. 126
Government Housing Loan Corporation (GHLC) 32, 36, 66, 79
Grandchildren 22, 23, 29, 38
Grandparents 8, 21, 22, 23, 24
Gross National Product (GNP) 31, 58, 59, 61

Hajime 84, 85, 122
Harper, S. 21, 22, 24, 52
Hashimoto, A. 12, 13, 14, 91
Helena 93, 107
Hideo 73, 79, 122
Hirai, Mrs. 93

153

Index

Hiraoka, K. 49, 59, 125, 126
Hirayama, Y. 26, 32, 35, 66, 81, 134
HIV/AIDS 23, 24
Home care 51, 55, 56, 60
Home ownership 3, 26, 27, 29, 30, 31, 32, 34, 35, 36, 66, 67, 68, 69, 70, 71, 72, 81, 86, 88, 89, 133
Hong Kong 31, 32, 36, 51, 54, 60, 132
House price indices 80
Housing assets 28, 36, 44, 45, 74, 75, 87, 89, 92, 93, 96, 99, 105, 106, 113, 134; *see also* Assets
Housing equity 34, 36, 37, 45, 107
Housing ladder 67, 68, 70, 77, 85, 133
Housing market 3, 26, 32, 33, 73, 80, 100, 107, 108, 133
Housing stock 33, 45, 46, 77, 84, 107

Ikegami, N. 57
Independent living 20, 51, 81, 82, 91, 112
In-depth interviews 65, 87
Individualism 2, 13
Inflation 3, 32, 33, 55, 68, 71, 86
Informal care 50, 51, 53, 56, 57, 60, 62, 63, 117, 118, 132, 133
Inheritance 1, 4, 28, 35, 36, 37, 38, 39, 41, 42, 43, 67, 70, 71, 83, 85, 88, 89, 91, 94, 95, 97, 99, 102, 110, 111, 113, 114, 122, 127, 129, 133, 134, 135
 Inheritance law 40, 43, 135
 Inheritance tax 29, 43, 44, 102, 134
Inter vivo 29, 40, 41, 44, 102, 134
Intergenerational relations 4, 13, 15, 16, 18, 28, 95, 106, 135; *see also* Family relations
Intergenerational transfer 26, 29, 110; *see also* Asset transfer; Property transfer
Investment 38, 39, 72, 74, 76, 77, 95, 96, 105
Izanagi Keiki [economy growth] 66
Izuhara, M. 4, 5, 12, 21, 28, 29, 35, 36, 38, 39, 41, 45, 46, 58, 77, 79, 88, 99, 112, 131

Jackie 75, 96, 116
Jennie 80

Keiko 104, 123
Kita, Mrs. 125
Koji 116

Korea 14, 15, 19, 31, 40, 41, 51, 52, 58, 131, 132, 133

Later life 45, 90, 97, 105
Legal reform 13, 14, 41
Life-course 6, 12, 19, 37, 38, 68, 84, 94, 95, 96, 106, 107, 110, 111, 118, 120, 124
Lloyd, L. 111, 117
Long-term care 47, 48, 49, 50, 51, 53, 56, 57, 60, 72, 111, 113, 132
Longevity 18, 37, 38, 101
Lost decade 134
Lowenstein, A. 16, 18, 19, 20
Low income 27, 58, 119

Mari 72, 121
Marriage 12, 13, 23, 40, 42, 51, 66, 75, 81, 97, 104
Marriage pattern 91
Mary 112, 113, 115
Masami 76, 118
Matsuo, Mrs. 92
McKenzie, Mrs. 80, 93
Meaning of home 27, 74, 88, 89, 90
Methods 5, 6, 7; *see also* Purposive sampling; Qualitative method
Michiko 79, 120
Migration/migrants 52, 67
Mika 115, 123
Ministry
 of Construction 33
 of Finance 43
 of Internal Affairs and Communications 71
 of Health and Welfare 66, 91, 104
 of Health, Labour and Welfare 97, 125
 of Land, Infrastructure and Transport 74, 85, 107
 of Telecommunication 89
Mixed economy of care 54, 56
Mixed economy of provision 56
Mixed economy of welfare 49, 66
Mobility 21, 24, 39, 52, 67, 70, 75
Morita, Mrs. 75
Munro, M. 4, 26, 42, 74

Nanri, Mrs. 113
Negative equity 26, 81
Negotiate family support 65, 99, 111, 112, 113, 130
Neoliberal 27, 105
Net provider 12

Net receiver 12
Netherlands (The) 42, 51, 57, 61, 62
New families 14
Nishi, Mr. 68, 71, 100, 101, 105
Nordic 30, 52, 53, 57, 58

Obedience 13, 14, 15
Occupational welfare *See* Company housing
OECD 50, 51, 53, 54, 57, 58, 60, 61, 62
Old age 11, 12, 28, 34, 44, 45, 92, 107, 133
Older people 2, 11, 18, 36, 40, 45, 48, 49, 51, 55, 66, 105, 106, 107, 111

Pam 80, 118, 120
Parasite singles 12, 81
Patriarchal/patriarchy 13, 14, 35, 38
Paul 77, 83, 117
Penny 77, 83, 95, 102, 121
Policy 2, 3, 5, 18, 27, 31, 47, 56; *see also* Social policy
Single Child Policy 19
Policy logic 2, 47, 53
Policy options 131
Policy reform 2, 49
Policy shift 112, 124
Porter, Mrs. 90, 92
Post-bubble 26, 33, 131
Prefab [prefabricated] 69
Privacy 90, 112
Privatization 30, 31, 62
Property ladder *see also* Housing ladder
Property transfer 29, 37, 88, 89, 93, 103, 104
Public housing *see* Council housing
Purposive sampling 6

Qualitative method 4
Quasi-market 55

Recession 33, 55, 81, 105
Reciprocity 1, 8, 9, 10, 12, 20, 21, 28, 45, 54, 110, 113, 114, 116, 118, 120, 123; *see also* Asymmetrical reciprocity
Recognition 62, 118, 119
Renting 45, 67, 69, 83, 84
Redistribution 10
Residential property 28, 43, 89, 90, 96, 101, 102, 104, 123, 129
Reward 3, 9, 27, 56, 62, 114, 115, 116, 118, 119, 120, 123
Right to Buy 26, 30, 36, 69

Ritual 14, 22, 75, 88
Rowlingson, K. 4, 38, 43, 44, 70, 71, 93, 95, 102, 110
Rural communities 14, 15, 16, 24, 52, 89, 110

Sally 99, 115, 117
Savings 27, 32, 37, 53, 89, 100, 105
Sayuri 124
Separation *See* Divorce
Sheltered housing 99, 107
Sibling 8, 39, 41, 42
Singapore 31, 32, 47, 54, 60, 132
Single-family home 34, 66, 74, 75
Social change 1, 17, 21, 100, 102, 124, 134, 135, 136
Social contract 1, 44, 134, 135
Social insurance 56, 57, 58, 111, 133
Social insurance on long-term care 2, 55, 62, 64, 117, 124, 131, 136
Social policy 36, 37, 44, 48, 54, 91, 115, 131, 132, 134
Social risk 48
Social status 13, 27
Social stratification 27, 46
Socialization of care 49, 58, 112, 117, 125
Soga, Mr. 71, 100, 126
Solidarity-conflict 16, 17, 20; *see also* Intergenerational Solidarity-Conflict model
Spain 22, 30, 31, 32, 51, 54, 60
State assistance 32
Stepfamilies 40, 97
Sustainability 64
Suzuki, Mrs. 78, 89
Symmetrical 10, 15; *see also* Exchange
Sweden 19, 40, 50, 54, 56, 58, 61

Taguchi, Mr. 67
Taiwan 15, 31, 54, 132
Tanaka, Mrs. 88
Tax concessions 47
Tenure choice 66, 74, 81, 83, 85, 133
Thailand 15, 19
Thompson, Mr. 70, 80
Thorns, D.C. 26, 27, 29, 35, 38, 74
Tokyo Women's Foundation 4, 35, 39, 43, 71
Toshiko 68, 96
Trading down 45, 107
Trajectories 32
  Housing market trajectories 80, 86

Index

Ungerson, C. 2, 52, 57, 60, 61, 62
US 10, 11, 12, 19, 20, 23, 29, 30, 32, 40

Walker, Alan 1, 4, 9, 12, 16, 47
Watanabe, Mr. 108
Wealth gap 27
Welfare regime 5, 24, 48, 53, 123, 132
Welfare state 2, 17, 27, 48, 53, 54, 55, 56, 60, 131
Will 41, 90, 92, 94, 95, 98, 99, 100, 119

Yamaguchi, Mrs. 106, 125
Yamashita, Mrs. 126
Yoko 128, 129
Yoshi [adopted husband] 75
Younger generation 2, 11, 16, 31, 38, 39, 44, 63, 76, 81, 85, 97, 99, 101, 111, 112, 114, 123
Younger people *see* Younger generation
Yuki 84, 85, 103, 122

Zena 115

For Product Safety Concerns and Information please contact our EU representative GPSR@taylorandfrancis.com
Taylor & Francis Verlag GmbH, Kaufingerstraße 24, 80331 München, Germany

www.ingramcontent.com/pod-product-compliance
Lightning Source LLC
Chambersburg PA
CBHW061841300426
44115CB00013B/2466